Practice

Eureka Math®
Grade 2 Fluency
Modules 6–8

Students, families, and educators:

Thank you for being part of the *Eureka Math*® community, where we celebrate the joy, wonder, and thrill of mathematics. One of the most obvious ways we display our excitement is through the fluency activities provided in *Eureka Math Practice*.

What is fluency in mathematics?

You may think of *fluency* as associated with the language arts, where it refers to speaking and writing with ease. In prekindergarten through grade 5, the *Eureka Math* curriculum contains multiple daily opportunities to build fluency *in mathematics*. Each is designed with the same notion—growing every student's ability to use mathematics *with ease*. Fluency experiences are generally fast-paced and energetic, celebrating improvement and focusing on recognizing patterns and connections within the material. They are not intended to be graded.

Eureka Math fluency activities provide differentiated practice through a variety of formats—some are conducted orally, some use manipulatives, others use a personal whiteboard, and still others use a handout and paper-and-pencil format. *Eureka Math Practice* provides each student with the printed fluency exercises for his or her grade level.

What is a Sprint?

Many printed fluency activities utilize the format we call a Sprint. These exercises build speed and accuracy with already acquired skills. Used when students are nearing optimum proficiency, Sprints leverage tempo to build a low-stakes adrenaline boost that increases memory and recall. Their intentional design makes Sprints inherently differentiated; the problems build from simple to complex, with the first quadrant of problems being the simplest and each subsequent quadrant adding complexity. Further, intentional patterns within the sequence of problems engage students' higher order thinking skills.

The suggested format for delivering a Sprint calls for students to do two consecutive Sprints (labeled A and B) on the same skill, each timed at one minute. Students pause between Sprints to articulate the patterns they noticed as they worked the first Sprint. Noticing the patterns often provides a natural boost to their performance on the second Sprint.

Sprints can be conducted with an untimed protocol as well. The untimed protocol is highly recommended when students are still building confidence with the level of complexity of the first quadrant of problems. Once all students are prepared for success on the Sprint, the work of improving speed and accuracy with the energy of a timed protocol is often welcome and invigorating.

Where can I find other fluency activities?

The *Eureka Math Teacher Edition* guides educators in the delivery of all fluency activities for each lesson, including those that do not require print materials. Additionally, the *Eureka Digital Suite* provides access to the fluency activities for all grade levels, searchable by standard or lesson.

Best wishes for a year filled with aha moments!

Jill Diniz

Jill Diniz
Director of Mathematics
Great Minds

Learn ◆ Practice ◆ Succeed

Eureka Math® student materials for *A Story of Units*® (K–5) are available in the *Learn, Practice, Succeed* trio. This series supports differentiation and remediation while keeping student materials organized and accessible. Educators will find that the *Learn, Practice,* and *Succeed* series also offers coherent—and therefore, more effective—resources for Response to Intervention (RTI), extra practice, and summer learning.

Learn

Eureka Math Learn serves as a student's in-class companion where they show their thinking, share what they know, and watch their knowledge build every day. *Learn* assembles the daily classwork—Application Problems, Exit Tickets, Problem Sets, templates—in an easily stored and navigated volume.

Practice

Each *Eureka Math* lesson begins with a series of energetic, joyous fluency activities, including those found in *Eureka Math Practice.* Students who are fluent in their math facts can master more material more deeply. With *Practice,* students build competence in newly acquired skills and reinforce previous learning in preparation for the next lesson.

Together, *Learn* and *Practice* provide all the print materials students will use for their core math instruction.

Succeed

Eureka Math Succeed enables students to work individually toward mastery. These additional problem sets align lesson by lesson with classroom instruction, making them ideal for use as homework or extra practice. Each problem set is accompanied by a Homework Helper, a set of worked examples that illustrate how to solve similar problems.

Teachers and tutors can use *Succeed* books from prior grade levels as curriculum-consistent tools for filling gaps in foundational knowledge. Students will thrive and progress more quickly as familiar models facilitate connections to their current grade-level content.

Contents

Module 6

Module 7

Module 8

Grade 2
Module 6

Name _____   Date _____

1.	10 + 3 =		21.	7 + 9 =
2.	10 + 6 =		22.	4 + 8 =
3.	10 + 4 =		23.	5 + 9 =
4.	5 + 10 =		24.	8 + 6 =
5.	8 + 10 =		25.	7 + 5 =
6.	10 + 9 =		26.	5 + 8 =
7.	12 + 2 =		27.	8 + 3 =
8.	13 + 4 =		28.	9 + 8 =
9.	16 + 3 =		29.	6 + 5 =
10.	2 + 17 =		30.	7 + 6 =
11.	5 + 14 =		31.	4 + 6 =
12.	7 + 12 =		32.	8 + 7 =
13.	16 + 3 =		33.	7 + 7 =
14.	11 + 5 =		34.	8 + 6 =
15.	9 + 2 =		35.	6 + 9 =
16.	5 + 9 =		36.	8 + 5 =
17.	7 + 9 =		37.	4 + 7 =
18.	9 + 4 =		38.	3 + 9 =
19.	7 + 8 =		39.	6 + 6 =
20.	8 + 8 =		40.	4 + 9 =

Lesson 1:   Use manipulatives to create equal groups.

© 2015 Great Minds®. eureka-math.org

3

Name _____   Date _____

1.	10 + 4 =		21.	4 + 8 =
2.	10 + 9 =		22.	7 + 6 =
3.	5 + 10 =		23.	_____ + 4 = 11
4.	2 + 10 =		24.	_____ + 8 = 13
5.	11 + 4 =		25.	6 + _____ = 14
6.	12 + 5 =		26.	8 + _____ = 15
7.	16 + 2 =		27.	_____ = 9 + 8
8.	13 + _____ = 18		28.	_____ = 4 + 7
9.	11 + _____ = 20		29.	_____ = 7 + 8
10.	14 + 3 =		30.	3 + 9 =
11.	_____ = 3 + 16		31.	6 + 7 =
12.	_____ = 7 + 12		32.	8 + _____ = 13
13.	_____ = 15 + 4		33.	_____ = 7 + 9
14.	9 + 2 =		34.	6 + 5 =
15.	6 + 9 =		35.	_____ = 5 + 7
16.	_____ + 4 = 11		36.	_____ = 8 + 4
17.	_____ + 6 = 13		37.	15 = 8 + _____
18.	_____ + 5 = 12		38.	17 = _____ + 9
19.	8 + 8 =		39.	14 = _____ + 7
20.	6 + 6 =		40.	19 = 8 + _____

Name _____   Date _____

1.	12 – 2 =	21.	16 – 9 =
2.	18 – 8 =	22.	14 – 6 =
3.	19 – 10 =	23.	16 – 8 =
4.	14 – 10 =	24.	15 – 6 =
5.	16 – 6 =	25.	17 – 8 =
6.	11 – 10 =	26.	18 – 9 =
7.	17 – 12 =	27.	15 – 7 =
8.	20 – 10 =	28.	13 – 8 =
9.	13 – 11 =	29.	11 – 3 =
10.	18 – 13 =	30.	12 – 5 =
11.	12 – 3 =	31.	11 – 2 =
12.	11 – 2 =	32.	13 – 6 =
13.	14 – 2 =	33.	16 – 7 =
14.	13 – 4 =	34.	12 – 8 =
15.	11 – 3 =	35.	16 – 13 =
16.	13 – 2 =	36.	15 – 14 =
17.	12 – 4 =	37.	17 – 12 =
18.	14 – 5 =	38.	19 – 16 =
19.	11 – 4 =	39.	18 – 11 =
20.	12 – 5 =	40.	20 – 16 =

Name _____   Date _____

1.	19 – 9 =	21.	16 – 7 =
2.	12 – 10 =	22.	17 – 8 =
3.	18 – 11 =	23.	16 – 7 =
4.	15 – 10 =	24.	14 – 8 =
5.	17 – 12 =	25.	17 – 9 =
6.	16 – 13 =	26.	12 – 9 =
7.	12 – 2 =	27.	16 – 8 =
8.	20 – 10 =	28.	15 – 7 =
9.	14 – 11 =	29.	13 – 8 =
10.	13 – 3 =	30.	14 – 7 =
11.	_____ = 11 – 3	31.	13 – 9 =
12.	_____ = 14 – 4	32.	15 – 9 =
13.	_____ = 13 – 4	33.	14 – 6 =
14.	_____ = 11 – 4	34.	_____ = 13 – 5
15.	_____ = 12 – 3	35.	_____ = 15 – 8
16.	_____ = 13 – 2	36.	_____ = 18 – 9
17.	_____ = 11 – 2	37.	_____ = 20 – 4
18.	16 – 8 =	38.	_____ = 20 – 17
19.	15 – 6 =	39.	_____ = 20 – 11
20.	12 – 5 =	40.	_____ = 20 – 3

Name _____   Date _____

1.	13 + 3 =	21.	11 – 8 =
2.	12 + 8 =	22.	13 – 7 =
3.	16 + 2 =	23.	15 – 8 =
4.	11 + 7 =	24.	12 + 6 =
5.	6 + 9 =	25.	13 + 2 =
6.	7 + 8 =	26.	9 + 11 =
7.	4 + 7 =	27.	6 + 8 =
8.	13 – 5 =	28.	8 + 9 =
9.	16 – 6 =	29.	7 + 5 =
10.	17 – 9 =	30.	13 – 7 =
11.	14 – 6 =	31.	15 – 8 =
12.	18 – 7 =	32.	11 – 9 =
13.	8 + 8 =	33.	12 – 3 =
14.	7 + 6 =	34.	14 – 5 =
15.	4 + 9 =	35.	13 + 6 =
16.	5 + 7 =	36.	8 + 5 =
17.	6 + 5 =	37.	4 + 7 =
18.	13 – 8 =	38.	7 + 8 =
19.	16 – 9 =	39.	4 + 9 =
20.	14 – 8 =	40.	20 – 12 =

A

Number Correct: _____

Subtraction Within 20

1.	11 − 10 =	
2.	12 − 10 =	
3.	13 − 10 =	
4.	19 − 10 =	
5.	11 − 1 =	
6.	12 − 2 =	
7.	13 − 3 =	
8.	17 − 7 =	
9.	11 − 2 =	
10.	11 − 3 =	
11.	11 − 4 =	
12.	11 − 8 =	
13.	18 − 8 =	
14.	13 − 4 =	
15.	13 − 5 =	
16.	13 − 6 =	
17.	13 − 8 =	
18.	16 − 6 =	
19.	12 − 3 =	
20.	12 − 4 =	
21.	12 − 5 =	
22.	12 − 9 =	

23.	19 − 9 =	
24.	15 − 6 =	
25.	15 − 7 =	
26.	15 − 9 =	
27.	20 − 10 =	
28.	14 − 5 =	
29.	14 − 6 =	
30.	14 − 7 =	
31.	14 − 9 =	
32.	15 − 5 =	
33.	17 − 8 =	
34.	17 − 9 =	
35.	18 − 8 =	
36.	16 − 7 =	
37.	16 − 8 =	
38.	16 − 9 =	
39.	17 − 10 =	
40.	12 − 8 =	
41.	18 − 9 =	
42.	11 − 9 =	
43.	15 − 8 =	
44.	13 − 7 =	

B

Number Correct: _____

Subtraction Within 20

Improvement: _____

1.	11 – 1 =		23.	16 – 6 =	
2.	12 – 2 =		24.	14 – 5 =	
3.	13 – 3 =		25.	14 – 6 =	
4.	18 – 8 =		26.	14 – 7 =	
5.	11 – 10 =		27.	14 – 9 =	
6.	12 – 10 =		28.	20 – 10 =	
7.	13 – 10 =		29.	15 – 6 =	
8.	18 – 10 =		30.	15 – 7 =	
9.	11 – 2 =		31.	15 – 9 =	
10.	11 – 3 =		32.	14 – 4 =	
11.	11 – 4 =		33.	16 – 7 =	
12.	11 – 7 =		34.	16 – 8 =	
13.	19 – 9 =		35.	16 – 9 =	
14.	12 – 3 =		36.	20 – 10 =	
15.	12 – 4 =		37.	17 – 8 =	
16.	12 – 5 =		38.	17 – 9 =	
17.	12 – 8 =		39.	16 – 10 =	
18.	17 – 7 =		40.	18 – 9 =	
19.	13 – 4 =		41.	12 – 9 =	
20.	13 – 5 =		42.	13 – 7 =	
21.	13 – 6 =		43.	11 – 8 =	
22.	13 – 9 =		44.	15 – 8 =	

Lesson 3:   Use math drawings to represent equal groups, and relate to repeated addition.

15

A

Number Correct: _____

Adding Crossing Ten

1.	9 + 1 =		23.	7 + 3 =	
2.	9 + 2 =		24.	7 + 4 =	
3.	9 + 3 =		25.	7 + 5 =	
4.	9 + 9 =		26.	7 + 9 =	
5.	8 + 2 =		27.	6 + 4 =	
6.	8 + 3 =		28.	6 + 5 =	
7.	8 + 4 =		29.	6 + 6 =	
8.	8 + 9 =		30.	6 + 9 =	
9.	9 + 1 =		31.	5 + 5 =	
10.	9 + 4 =		32.	5 + 6 =	
11.	9 + 5 =		33.	5 + 7 =	
12.	9 + 8 =		34.	5 + 9 =	
13.	8 + 2 =		35.	4 + 6 =	
14.	8 + 5 =		36.	4 + 7 =	
15.	8 + 6 =		37.	4 + 9 =	
16.	8 + 8 =		38.	3 + 7 =	
17.	9 + 1 =		39.	3 + 9 =	
18.	9 + 7 =		40.	5 + 8 =	
19.	8 + 2 =		41.	2 + 8 =	
20.	8 + 7 =		42.	4 + 8 =	
21.	9 + 1 =		43.	1 + 9 =	
22.	9 + 6 =		44.	2 + 9 =	

EUREKA MATH®

Lesson 4:   Represent equal groups with tape diagrams, and relate to repeated addition.

17

B   Number Correct: _____

Adding Crossing Ten   Improvement: _____

1.	8 + 2 =		23.	7 + 3 =	
2.	8 + 3 =		24.	7 + 4 =	
3.	8 + 4 =		25.	7 + 5 =	
4.	8 + 8 =		26.	7 + 8 =	
5.	9 + 1 =		27.	6 + 4 =	
6.	9 + 2 =		28.	6 + 5 =	
7.	9 + 3 =		29.	6 + 6 =	
8.	9 + 8 =		30.	6 + 8 =	
9.	8 + 2 =		31.	5 + 5 =	
10.	8 + 5 =		32.	5 + 6 =	
11.	8 + 6 =		33.	5 + 7 =	
12.	8 + 9 =		34.	5 + 8 =	
13.	9 + 1 =		35.	4 + 6 =	
14.	9 + 4 =		36.	4 + 7 =	
15.	9 + 5 =		37.	4 + 8 =	
16.	9 + 9 =		38.	3 + 7 =	
17.	9 + 1 =		39.	3 + 9 =	
18.	9 + 7 =		40.	5 + 9 =	
19.	8 + 2 =		41.	2 + 8 =	
20.	8 + 7 =		42.	4 + 9 =	
21.	9 + 1 =		43.	1 + 9 =	
22.	9 + 6 =		44.	2 + 9 =	

EUREKA
MATH

Lesson 4:   Represent equal groups with tape diagrams, and relate to repeated
              addition.

© 2015 Great Minds®. eureka-math.org

19

A

Number Correct: _____

Sums to the Teens

1.	9 + 2 =		23.	4 + 7 =	
2.	9 + 3 =		24.	4 + 8 =	
3.	9 + 4 =		25.	5 + 6 =	
4.	9 + 7 =		26.	5 + 7 =	
5.	7 + 9 =		27.	3 + 8 =	
6.	10 + 1 =		28.	3 + 9 =	
7.	10 + 2 =		29.	2 + 9 =	
8.	10 + 3 =		30.	5 + 10 =	
9.	10 + 8 =		31.	5 + 8 =	
10.	8 + 10 =		32.	9 + 6 =	
11.	8 + 3 =		33.	6 + 9 =	
12.	8 + 4 =		34.	7 + 6 =	
13.	8 + 5 =		35.	6 + 7 =	
14.	8 + 9 =		36.	8 + 6 =	
15.	9 + 8 =		37.	6 + 8 =	
16.	7 + 4 =		38.	8 + 7 =	
17.	10 + 5 =		39.	7 + 8 =	
18.	6 + 5 =		40.	6 + 6 =	
19.	7 + 5 =		41.	7 + 7 =	
20.	9 + 5 =		42.	8 + 8 =	
21.	5 + 9 =		43.	9 + 9 =	
22.	10 + 6 =		44.	4 + 9 =	

EUREKA
MATH®

Lesson 7:   Represent arrays and distinguish rows and columns using math
             drawings.

21

© 2015 Great Minds®. eureka-math.org

B

Number Correct: _____

Improvement: _____

Sums to the Teens

1.	10 + 1 =	
2.	10 + 2 =	
3.	10 + 3 =	
4.	10 + 9 =	
5.	9 + 10 =	
6.	9 + 2 =	
7.	9 + 3 =	
8.	9 + 4 =	
9.	9 + 8 =	
10.	8 + 9 =	
11.	8 + 3 =	
12.	8 + 4 =	
13.	8 + 5 =	
14.	8 + 7 =	
15.	7 + 8 =	
16.	7 + 4 =	
17.	10 + 4 =	
18.	6 + 5 =	
19.	7 + 5 =	
20.	9 + 5 =	
21.	5 + 9 =	
22.	10 + 8 =	

23.	5 + 6 =	
24.	5 + 7 =	
25.	4 + 7 =	
26.	4 + 8 =	
27.	4 + 10 =	
28.	3 + 8 =	
29.	3 + 9 =	
30.	2 + 9 =	
31.	5 + 8 =	
32.	7 + 6 =	
33.	6 + 7 =	
34.	8 + 6 =	
35.	6 + 8 =	
36.	9 + 6 =	
37.	6 + 9 =	
38.	9 + 7 =	
39.	7 + 9 =	
40.	6 + 6 =	
41.	7 + 7 =	
42.	8 + 8 =	
43.	9 + 9 =	
44.	4 + 9 =	

EUREKA MATH

Lesson 7: Represent arrays and distinguish rows and columns using math drawings.

23

A

Number Correct: _____

Subtraction from Teens

1.	11 – 10 =		23.	19 – 9 =	
2.	12 – 10 =		24.	15 – 6 =	
3.	13 – 10 =		25.	15 – 7 =	
4.	19 – 10 =		26.	15 – 9 =	
5.	11 – 1 =		27.	20 – 10 =	
6.	12 – 2 =		28.	14 – 5 =	
7.	13 – 3 =		29.	14 – 6 =	
8.	17 – 7 =		30.	14 – 7 =	
9.	11 – 2 =		31.	14 – 9 =	
10.	11 – 3 =		32.	15 – 5 =	
11.	11 – 4 =		33.	17 – 8 =	
12.	11 – 8 =		34.	17 – 9 =	
13.	18 – 8 =		35.	18 – 8 =	
14.	13 – 4 =		36.	16 – 7 =	
15.	13 – 5 =		37.	16 – 8 =	
16.	13 – 6 =		38.	16 – 9 =	
17.	13 – 8 =		39.	17 – 10 =	
18.	16 – 6 =		40.	12 – 8 =	
19.	12 – 3 =		41.	18 – 9 =	
20.	12 – 4 =		42.	11 – 9 =	
21.	12 – 5 =		43.	15 – 8 =	
22.	12 – 9 =		44.	13 – 7 =	

EUREKA MATH

Lesson 8: Create arrays using square tiles with gaps.

25

B

Number Correct: _____

Subtraction from Teens

Improvement: _____

1.	11 – 1 =	
2.	12 – 2 =	
3.	13 – 3 =	
4.	18 – 8 =	
5.	11 – 10 =	
6.	12 – 10 =	
7.	13 – 10 =	
8.	18 – 10 =	
9.	11 – 2 =	
10.	11 – 3 =	
11.	11 – 4 =	
12.	11 – 7 =	
13.	19 – 9 =	
14.	12 – 3 =	
15.	12 – 4 =	
16.	12 – 5 =	
17.	12 – 8 =	
18.	17 – 7 =	
19.	13 – 4 =	
20.	13 – 5 =	
21.	13 – 6 =	
22.	13 – 9 =	

23.	16 – 6 =	
24.	14 – 5 =	
25.	14 – 6 =	
26.	14 – 7 =	
27.	14 – 9 =	
28.	20 – 10 =	
29.	15 – 6 =	
30.	15 – 7 =	
31.	15 – 9 =	
32.	14 – 4 =	
33.	16 – 7 =	
34.	16 – 8 =	
35.	16 – 9 =	
36.	20 – 10 =	
37.	17 – 8 =	
38.	17 – 9 =	
39.	16 – 10 =	
40.	18 – 9 =	
41.	12 – 9 =	
42.	13 – 7 =	
43.	11 – 8 =	
44.	15 – 8 =	

A

Number Correct: _____

Sums to the Teens

1.	9 + 1 =		23.	7 + 3 =	
2.	9 + 2 =		24.	7 + 4 =	
3.	9 + 3 =		25.	7 + 5 =	
4.	9 + 9 =		26.	7 + 9 =	
5.	8 + 2 =		27.	6 + 4 =	
6.	8 + 3 =		28.	6 + 5 =	
7.	8 + 4 =		29.	6 + 6 =	
8.	8 + 9 =		30.	6 + 9 =	
9.	9 + 1 =		31.	5 + 5 =	
10.	9 + 4 =		32.	5 + 6 =	
11.	9 + 5 =		33.	5 + 7 =	
12.	9 + 8 =		34.	5 + 9 =	
13.	8 + 2 =		35.	4 + 6 =	
14.	8 + 5 =		36.	4 + 7 =	
15.	8 + 6 =		37.	4 + 9 =	
16.	8 + 8 =		38.	3 + 7 =	
17.	9 + 1 =		39.	3 + 9 =	
18.	9 + 7 =		40.	5 + 8 =	
19.	8 + 2 =		41.	2 + 8 =	
20.	8 + 7 =		42.	4 + 8 =	
21.	9 + 1 =		43.	1 + 9 =	
22.	9 + 6 =		44.	2 + 9 =	

Lesson 10:   Use square tiles to compose a rectangle, and relate to the array model.

29

B

Number Correct: _____

Sums to the Teens

Improvement: _____

1.	8 + 2 =	
2.	8 + 3 =	
3.	8 + 4 =	
4.	8 + 8 =	
5.	9 + 1 =	
6.	9 + 2 =	
7.	9 + 3 =	
8.	9 + 8 =	
9.	8 + 2 =	
10.	8 + 5 =	
11.	8 + 6 =	
12.	8 + 9 =	
13.	9 + 1 =	
14.	9 + 4 =	
15.	9 + 5 =	
16.	9 + 9 =	
17.	9 + 1 =	
18.	9 + 7 =	
19.	8 + 2 =	
20.	8 + 7 =	
21.	9 + 1 =	
22.	9 + 6 =	

23.	7 + 3 =	
24.	7 + 4 =	
25.	7 + 5 =	
26.	7 + 8 =	
27.	6 + 4 =	
28.	6 + 5 =	
29.	6 + 6 =	
30.	6 + 8 =	
31.	5 + 5 =	
32.	5 + 6 =	
33.	5 + 7 =	
34.	5 + 8 =	
35.	4 + 6 =	
36.	4 + 7 =	
37.	4 + 8 =	
38.	3 + 7 =	
39.	3 + 9 =	
40.	5 + 9 =	
41.	2 + 8 =	
42.	4 + 9 =	
43.	1 + 9 =	
44.	2 + 9 =	

A

Number Correct: _____

Subtraction Crossing Ten

1.	10 – 5 =	
2.	20 – 5 =	
3.	30 – 5 =	
4.	10 – 2 =	
5.	20 – 2 =	
6.	30 – 2 =	
7.	11 – 2 =	
8.	21 – 2 =	
9.	31 – 2 =	
10.	10 – 8 =	
11.	11 – 8 =	
12.	21 – 8 =	
13.	31 – 8 =	
14.	14 – 5 =	
15.	24 – 5 =	
16.	34 – 5 =	
17.	15 – 6 =	
18.	25 – 6 =	
19.	35 – 6 =	
20.	10 – 7 =	
21.	20 – 8 =	
22.	30 – 9 =	

23.	14 – 6 =	
24.	24 – 6 =	
25.	34 – 6 =	
26.	15 – 7 =	
27.	25 – 7 =	
28.	35 – 7 =	
29.	11 – 4 =	
30.	21 – 4 =	
31.	31 – 4 =	
32.	12 – 6 =	
33.	22 – 6 =	
34.	32 – 6 =	
35.	21 – 6 =	
36.	31 – 6 =	
37.	12 – 8 =	
38.	32 – 8 =	
39.	21 – 8 =	
40.	31 – 8 =	
41.	28 – 9 =	
42.	27 – 8 =	
43.	38 – 9 =	
44.	37 – 8 =	

B

Number Correct: _____

Improvement: _____

Subtraction Crossing Ten

1.	10 – 1 =	
2.	20 – 1 =	
3.	30 – 1 =	
4.	10 – 3 =	
5.	20 – 3 =	
6.	30 – 3 =	
7.	12 – 3 =	
8.	22 – 3 =	
9.	32 – 3 =	
10.	10 – 9 =	
11.	11 – 9 =	
12.	21 – 9 =	
13.	31 – 9 =	
14.	13 – 4 =	
15.	23 – 4 =	
16.	33 – 4 =	
17.	16 – 7 =	
18.	26 – 7 =	
19.	36 – 7 =	
20.	10 – 6 =	
21.	20 – 7 =	
22.	30 – 8 =	

23.	13 – 5 =	
24.	23 – 5 =	
25.	33 – 5 =	
26.	16 – 8 =	
27.	26 – 8 =	
28.	36 – 8 =	
29.	12 – 5 =	
30.	22 – 5 =	
31.	32 – 5 =	
32.	11 – 5 =	
33.	21 – 5 =	
34.	31 – 5 =	
35.	12 – 7 =	
36.	22 – 7 =	
37.	11 – 7 =	
38.	31 – 7 =	
39.	22 – 9 =	
40.	32 – 9 =	
41.	38 – 9 =	
42.	37 – 8 =	
43.	28 – 9 =	
44.	27 – 8 =	

EUREKA MATH®

Lesson 11: Use square tiles to compose a rectangle, and relate to the array model.

35

Name _____   Date _____

1.	10 + 2 =	21.	7 + 9 =
2.	10 + 7 =	22.	5 + 8 =
3.	10 + 5 =	23.	3 + 9 =
4.	4 + 10 =	24.	8 + 6 =
5.	6 + 11 =	25.	7 + 4 =
6.	12 + 2 =	26.	9 + 5 =
7.	14 + 3 =	27.	6 + 6 =
8.	13 + 5 =	28.	8 + 3 =
9.	17 + 2 =	29.	7 + 6 =
10.	12 + 6 =	30.	6 + 9 =
11.	11 + 9 =	31.	8 + 7 =
12.	2 + 16 =	32.	9 + 9 =
13.	15 + 4 =	33.	5 + 7 =
14.	5 + 9 =	34.	8 + 4 =
15.	9 + 2 =	35.	6 + 5 =
16.	4 + 9 =	36.	9 + 7 =
17.	9 + 6 =	37.	6 + 8 =
18.	8 + 9 =	38.	2 + 9 =
19.	7 + 8 =	39.	9 + 8 =
20.	8 + 8 =	40.	7 + 7 =

Lesson 12:   Use math drawings to compose a rectangle with square tiles.

Name _____   Date _____

1.	10 + 6 =	21.	3 + 8 =
2.	10 + 9 =	22.	9 + 4 =
3.	7 + 10 =	23.	_____ + 6 = 11
4.	3 + 10 =	24.	_____ + 9 = 13
5.	5 + 11 =	25.	8 + _____ = 14
6.	12 + 8 =	26.	7 + _____ = 15
7.	14 + 3 =	27.	_____ = 4 + 8
8.	13 + _____ = 19	28.	_____ = 8 + 9
9.	15 + _____ = 18	29.	_____ = 6 + 4
10.	12 + 5 =	30.	3 + 9 =
11.	_____ = 2 + 17	31.	5 + 7 =
12.	_____ = 3 + 13	32.	8 + _____ =14
13.	_____ = 16 + 2	33.	_____ = 5 + 9
14.	9 + 3 =	34.	8 + 8 =
15.	6 + 9 =	35.	_____ = 7 + 9
16.	_____ + 5 = 14	36.	_____ = 8 + 4
17.	_____ + 7 = 13	37.	17 = 8 + _____
18.	_____ + 8 = 12	38.	19 = _____ + 9
19.	8 + 7 =	39.	12 = _____ + 7
20.	7 + 6 =	40.	15 = 8 + _____

Name _____   Date _____

1.	13 – 3 =	21.	16 – 8 =
2.	19 – 9 =	22.	14 – 5 =
3.	15 – 10 =	23.	16 – 7 =
4.	18 – 10 =	24.	15 – 7 =
5.	12 – 2 =	25.	17 – 8 =
6.	11 – 10 =	26.	18 – 9 =
7.	17 – 13 =	27.	15 – 6 =
8.	20 – 10 =	28.	13 – 8 =
9.	14 – 11 =	29.	14 – 6 =
10.	16 – 12 =	30.	12 – 5 =
11.	11 – 3 =	31.	11 – 7 =
12.	13 – 2 =	32.	13 – 8 =
13.	14 – 2 =	33.	16 – 9 =
14.	13 – 4 =	34.	12 – 8 =
15.	12 – 3 =	35.	16 – 12 =
16.	11 – 4 =	36.	18 – 15 =
17.	12 – 5 =	37.	15 – 14 =
18.	14 – 5 =	38.	17 – 11 =
19.	11 – 2 =	39.	19 – 13 =
20.	12 – 4 =	40.	20 – 12 =

Name _____   Date _____

1.	$17 - 7 =$	21.	$16 - 7 =$
2.	$14 - 10 =$	22.	$17 - 8 =$
3.	$19 - 11 =$	23.	$18 - 7 =$
4.	$16 - 10 =$	24.	$14 - 6 =$
5.	$17 - 12 =$	25.	$17 - 8 =$
6.	$15 - 13 =$	26.	$12 - 8 =$
7.	$12 - 3 =$	27.	$14 - 7 =$
8.	$20 - 11 =$	28.	$15 - 8 =$
9.	$18 - 11 =$	29.	$13 - 5 =$
10.	$13 - 5 =$	30.	$16 - 8 =$
11.	_____ $= 11 - 2$	31.	$14 - 9 =$
12.	_____ $= 12 - 4$	32.	$15 - 6 =$
13.	_____ $= 13 - 5$	33.	$13 - 6 =$
14.	_____ $= 12 - 3$	34.	_____ $= 13 - 8$
15.	_____ $= 11 - 4$	35.	_____ $= 15 - 7$
16.	_____ $= 13 - 2$	36.	_____ $= 18 - 9$
17.	_____ $= 11 - 3$	37.	_____ $= 20 - 14$
18.	$17 - 8 =$	38.	_____ $= 20 - 7$
19.	$14 - 6 =$	39.	_____ $= 20 - 11$
20.	$16 - 9 =$	40.	_____ $= 20 - 8$

Name _____   Date _____

1.	$11 + 9 =$		21.	$13 - 7 =$
2.	$13 + 5 =$		22.	$11 - 8 =$
3.	$14 + 3 =$		23.	$15 - 6 =$
4.	$12 + 7 =$		24.	$12 + 7 =$
5.	$5 + 9 =$		25.	$14 + 3 =$
6.	$8 + 8 =$		26.	$8 + 12 =$
7.	$14 - 7 =$		27.	$5 + 7 =$
8.	$13 - 5 =$		28.	$8 + 9 =$
9.	$16 - 7 =$		29.	$7 + 5 =$
10.	$17 - 9 =$		30.	$13 - 6 =$
11.	$14 - 6 =$		31.	$14 - 8 =$
12.	$18 - 5 =$		32.	$12 - 9 =$
13.	$9 + 9 =$		33.	$11 - 3 =$
14.	$7 + 6 =$		34.	$14 - 5 =$
15.	$3 + 9 =$		35.	$13 - 8 =$
16.	$6 + 7 =$		36.	$8 + 5 =$
17.	$8 + 5 =$		37.	$4 + 7 =$
18.	$13 - 8 =$		38.	$7 + 8 =$
19.	$16 - 9 =$		39.	$4 + 9 =$
20.	$14 - 8 =$		40.	$20 - 8 =$

A

Number Correct: _____

Subtraction from Teens

1.	11 – 10 =	
2.	12 – 10 =	
3.	13 – 10 =	
4.	19 – 10 =	
5.	11 – 1 =	
6.	12 – 2 =	
7.	13 – 3 =	
8.	17 – 7 =	
9.	11 – 2 =	
10.	11 – 3 =	
11.	11 – 4 =	
12.	11 – 8 =	
13.	18 – 8 =	
14.	13 – 4 =	
15.	13 – 5 =	
16.	13 – 6 =	
17.	13 – 8 =	
18.	16 – 6 =	
19.	12 – 3 =	
20.	12 – 4 =	
21.	12 – 5 =	
22.	12 – 9 =	

23.	19 – 9 =	
24.	15 – 6 =	
25.	15 – 7 =	
26.	15 – 9 =	
27.	20 – 10 =	
28.	14 – 5 =	
29.	14 – 6 =	
30.	14 – 7 =	
31.	14 – 9 =	
32.	15 – 5 =	
33.	17 – 8 =	
34.	17 – 9 =	
35.	18 – 8 =	
36.	16 – 7 =	
37.	16 – 8 =	
38.	16 – 9 =	
39.	17 – 10 =	
40.	12 – 8 =	
41.	18 – 9 =	
42.	11 – 9 =	
43.	15 – 8 =	
44.	13 – 7 =	

EUREKA MATH®

Lesson 14:   Use scissors to partition a rectangle into same-size squares, and
compose arrays with the squares.

47

© 2015 Great Minds®. eureka-math.org

B

Number Correct: _____

Subtraction from Teens

Improvement: _____

1.	11 – 1 =	
2.	12 – 2 =	
3.	13 – 3 =	
4.	18 – 8 =	
5.	11 – 10 =	
6.	12 – 10 =	
7.	13 – 10 =	
8.	18 – 10 =	
9.	11 – 2 =	
10.	11 – 3 =	
11.	11 – 4 =	
12.	11 – 7 =	
13.	19 – 9 =	
14.	12 – 3 =	
15.	12 – 4 =	
16.	12 – 5 =	
17.	12 – 8 =	
18.	17 – 7 =	
19.	13 – 4 =	
20.	13 – 5 =	
21.	13 – 6 =	
22.	13 – 9 =	

23.	16 – 6 =	
24.	14 – 5 =	
25.	14 – 6 =	
26.	14 – 7 =	
27.	14 – 9 =	
28.	20 – 10 =	
29.	15 – 6 =	
30.	15 – 7 =	
31.	15 – 9 =	
32.	14 – 4 =	
33.	16 – 7 =	
34.	16 – 8 =	
35.	16 – 9 =	
36.	20 – 10 =	
37.	17 – 8 =	
38.	17 – 9 =	
39.	16 – 10 =	
40.	18 – 9 =	
41.	12 – 9 =	
42.	13 – 7 =	
43.	11 – 8 =	
44.	15 – 8 =	

EUREKA
MATH®

Lesson 14:   Use scissors to partition a rectangle into same-size squares, and
             compose arrays with the squares.

© 2015 Great Minds®. eureka-math.org

49

A

Number Correct: _____

Subtract Crossing the Ten

1.	10 – 1 =		23.	21 – 6 =	
2.	10 – 2 =		24.	91 – 6 =	
3.	20 – 2 =		25.	10 – 7 =	
4.	40 – 2 =		26.	11 – 7 =	
5.	10 – 2 =		27.	31 – 7 =	
6.	11 – 2 =		28.	10 – 8 =	
7.	21 – 2 =		29.	11 – 8 =	
8.	51 – 2=		30.	41 – 8 =	
9.	10 – 3 =		31.	10 – 9 =	
10.	11 – 3 =		32.	11 – 9 =	
11.	21 – 3 =		33.	51 – 9 =	
12.	61 – 3 =		34.	12 – 3 =	
13.	10 – 4 =		35.	82 – 3 =	
14.	11 – 4 =		36.	13 – 5 =	
15.	21 – 4 =		37.	73 – 5 =	
16.	71 – 4 =		38.	14 – 6 =	
17.	10 – 5 =		39.	84 – 6 =	
18.	11 – 5 =		40.	15 – 8 =	
19.	21 – 5 =		41.	95 – 8 =	
20.	81 – 5 =		42.	16 – 7 =	
21.	10 – 6 =		43.	46 – 7 =	
22.	11 – 6 =		44.	68 – 9 =	

EUREKA MATH

Lesson 15:   Use math drawings to partition a rectangle with square tiles, and
relate to repeated addition.

© 2015 Great Minds®. eureka-math.org

51

B

Subtract Crossing the Ten

1.	10 – 2 =	
2.	20 – 2 =	
3.	30 – 2 =	
4.	50 – 2 =	
5.	10 – 2 =	
6.	11 – 2 =	
7.	21 – 2 =	
8.	61 – 2 =	
9.	10 – 3 =	
10.	11 – 3 =	
11.	21 – 3 =	
12.	71 – 3 =	
13.	10 – 4 =	
14.	11 – 4 =	
15.	21 – 4 =	
16.	81 – 4 =	
17.	10 – 5 =	
18.	11 – 5 =	
19.	21 – 5 =	
20.	91 – 5 =	
21.	10 – 6 =	
22.	11 – 6 =	

23.	21 – 6 =	
24.	41 – 6 =	
25.	10 – 7 =	
26.	11 – 7 =	
27.	51 – 7 =	
28.	10 – 8 =	
29.	11 – 8 =	
30.	61 – 8 =	
31.	10 – 9 =	
32.	11 – 9 =	
33.	31 – 9 =	
34.	12 – 3 =	
35.	92 – 3 =	
36.	13 – 5 =	
37.	43 – 5 =	
38.	14 – 6 =	
39.	64 – 6 =	
40.	15 – 8 =	
41.	85 – 8 =	
42.	16 – 7 =	
43.	76 – 7 =	
44.	58 – 9 =	

Lesson 15:   Use math drawings to partition a rectangle with square tiles, and relate to repeated addition.

53

© 2015 Great Minds®. eureka-math.org

A

Number Correct: _____

Subtraction from Teens

1.	10 – 3 =	
2.	11 – 3 =	
3.	12 – 3 =	
4.	10 – 2 =	
5.	11 – 2 =	
6.	10 – 5 =	
7.	11 – 5 =	
8.	12 – 5 =	
9.	14 – 5 =	
10.	10 – 4 =	
11.	11 – 4 =	
12.	12 – 4 =	
13.	13 – 4 =	
14.	10 – 7 =	
15.	11 – 7 =	
16.	12 – 7 =	
17.	15 – 7 =	
18.	10 – 6 =	
19.	11 – 6 =	
20.	12 – 6 =	
21.	14 – 6 =	
22.	10 – 9 =	

23.	11 – 9 =	
24.	12 – 9 =	
25.	17 – 9 =	
26.	10 – 8 =	
27.	11 – 8 =	
28.	12 – 8 =	
29.	16 – 8 =	
30.	10 – 6 =	
31.	13 – 6 =	
32.	15 – 6 =	
33.	10 – 7 =	
34.	13 – 7 =	
35.	14 – 7 =	
36.	16 – 7 =	
37.	10 – 8 =	
38.	13 – 8 =	
39.	14 – 8 =	
40.	17 – 8 =	
41.	10 – 9 =	
42.	13 – 9 =	
43.	14 – 9 =	
44.	18 – 9 =	

EUREKA MATH

Lesson 18:   Pair objects and skip-count to relate to even numbers.

55

© 2015 Great Minds®. eureka-math.org

B

Number Correct: _____

Improvement: _____

Subtraction from Teens

1.	10 – 2 =	
2.	11 – 2 =	
3.	10 – 4 =	
4.	11 – 4 =	
5.	12 – 4 =	
6.	13 – 4 =	
7.	10 – 3 =	
8.	11 – 3 =	
9.	12 – 3 =	
10.	10 – 6 =	
11.	11 – 6 =	
12.	12 – 6 =	
13.	15 – 6 =	
14.	10 – 5 =	
15.	11 – 5 =	
16.	12 – 5 =	
17.	14 – 5 =	
18.	10 – 8 =	
19.	11 – 8 =	
20.	12 – 8 =	
21.	17 – 8 =	
22.	10 – 7 =	

23.	11 – 7 =	
24.	12 – 7 =	
25.	16 – 7 =	
26.	10 – 9 =	
27.	11 – 9 =	
28.	12 – 9 =	
29.	18 – 9 =	
30.	10 – 5 =	
31.	13 – 5 =	
32.	10 – 6 =	
33.	13 – 6 =	
34.	14 – 6 =	
35.	10 – 7 =	
36.	13 – 7 =	
37.	15 – 7 =	
38.	10 – 8 =	
39.	13 – 8 =	
40.	14 – 8 =	
41.	16 – 8 =	
42.	10 – 9 =	
43.	16 – 9 =	
44.	17 – 9 =	

Lesson 18:   Pair objects and skip-count to relate to even numbers.

EUREKA MATH

A

Number Correct: _____

Sums to the Teens

1.	9 + 2 =		23.	4 + 7 =	
2.	9 + 3 =		24.	4 + 8 =	
3.	9 + 4 =		25.	5 + 6 =	
4.	9 + 7 =		26.	5 + 7 =	
5.	7 + 9 =		27.	3 + 8 =	
6.	10 + 1 =		28.	3 + 9 =	
7.	10 + 2 =		29.	2 + 9 =	
8.	10 + 3 =		30.	5 + 10 =	
9.	10 + 8 =		31.	5 + 8 =	
10.	8 + 10 =		32.	9 + 6 =	
11.	8 + 3 =		33.	6 + 9 =	
12.	8 + 4 =		34.	7 + 6 =	
13.	8 + 5 =		35.	6 + 7 =	
14.	8 + 9 =		36.	8 + 6 =	
15.	9 + 8 =		37.	6 + 8 =	
16.	7 + 4 =		38.	8 + 7 =	
17.	10 + 5 =		39.	7 + 8 =	
18.	6 + 5 =		40.	6 + 6 =	
19.	7 + 5 =		41.	7 + 7 =	
20.	9 + 5 =		42.	8 + 8 =	
21.	5 + 9 =		43.	9 + 9 =	
22.	10 + 6 =		44.	4 + 9 =	

EUREKA
MATH®

Lesson 19: Investigate the pattern of even numbers: 0, 2, 4, 6, and 8 in the ones place, and relate to odd numbers.

© 2015 Great Minds®. eureka-math.org

59

B

Number Correct: _____

Improvement: _____

Sums to the Teens

1.	10 + 1 =	
2.	10 + 2 =	
3.	10 + 3 =	
4.	10 + 9 =	
5.	9 + 10 =	
6.	9 + 2 =	
7.	9 + 3 =	
8.	9 + 4 =	
9.	9 + 8 =	
10.	8 + 9 =	
11.	8 + 3 =	
12.	8 + 4 =	
13.	8 + 5 =	
14.	8 + 7 =	
15.	7 + 8 =	
16.	7 + 4 =	
17.	10 + 4 =	
18.	6 + 5 =	
19.	7 + 5 =	
20.	9 + 5 =	
21.	5 + 9 =	
22.	10 + 8 =	

23.	5 + 6 =	
24.	5 + 7 =	
25.	4 + 7 =	
26.	4 + 8 =	
27.	4 + 10 =	
28.	3 + 8 =	
29.	3 + 9 =	
30.	2 + 9 =	
31.	5 + 8 =	
32.	7 + 6 =	
33.	6 + 7 =	
34.	8 + 6 =	
35.	6 + 8 =	
36.	9 + 6 =	
37.	6 + 9 =	
38.	9 + 7 =	
39.	7 + 9 =	
40.	6 + 6 =	
41.	7 + 7 =	
42.	8 + 8 =	
43.	9 + 9 =	
44.	4 + 9 =	

EUREKA MATH

Lesson 19: Investigate the pattern of even numbers: 0, 2, 4, 6, and 8 in the ones place, and relate to odd numbers.

61

Grade 2
Module 7

Name _____   Date _____

1.	10 + 2 =	21.	7 + 9 =
2.	10 + 7 =	22.	5 + 8 =
3.	10 + 5 =	23.	3 + 9 =
4.	4 + 10 =	24.	8 + 6 =
5.	6 + 11 =	25.	7 + 4 =
6.	12 + 2 =	26.	9 + 5 =
7.	14 + 3 =	27.	6 + 6 =
8.	13 + 5 =	28.	8 + 3 =
9.	17 + 2 =	29.	7 + 6 =
10.	12 + 6 =	30.	6 + 9 =
11.	11 + 9 =	31.	8 + 7 =
12.	2 + 16 =	32.	9 + 9 =
13.	15 + 4 =	33.	5 + 7 =
14.	5 + 9 =	34.	8 + 4 =
15.	9 + 2 =	35.	6 + 5 =
16.	4 + 9 =	36.	9 + 7 =
17.	9 + 6 =	37.	6 + 8 =
18.	8 + 9 =	38.	2 + 9 =
19	7 + 8 =	39.	9 + 8 =
20.	8 + 8 =	40.	7 + 7 =

EUREKA
MATH

Lesson 1:   Sort and record data into a table using up to four categories; use
            category counts to solve word problems.

© 2015 Great Minds®. eureka-math.org

65

Name _____   Date _____

1.	10 + 6 =	21.	3 + 8 =
2.	10 + 9 =	22.	9 + 4 =
3.	7 + 10 =	23.	_____ + 6 = 11
4.	3 + 10 =	24.	_____ + 9 = 13
5.	5 + 11 =	25.	8 + _____ = 14
6.	12 + 8 =	26.	7 + _____ = 15
7.	14 + 3 =	27.	_____ = 4 + 8
8.	13 + _____ = 19	28.	_____ = 8 + 9
9.	15 + _____ = 18	29.	_____ = 6 + 4
10.	12 + 5 =	30.	3 + 9 =
11.	_____ = 2 + 17	31.	5 + 7 =
12.	_____ = 3 + 13	32.	8 + _____ = 14
13.	_____ = 16 + 2	33.	_____ = 5 + 9
14.	9 + 3 =	34.	8 + 8 =
15.	6 + 9 =	35.	_____ = 7 + 9
16.	_____ + 5 = 14	36.	_____ = 8 + 4
17.	_____ + 7 = 13	37.	17 = 8 + _____
18.	_____ + 8 = 12	38.	19 = _____ + 9
19	8 + 7 =	39.	12 = _____ + 7
20.	7 + 6 =	40.	15 = 8 + _____

EUREKA MATH   Lesson 1:   Sort and record data into a table using up to four categories; use category counts to solve word problems.   67

© 2015 Great Minds®. eureka-math.org

Name _____   Date _____

1.	13 – 3 =	21.	16 – 8 =
2.	19 – 9 =	22.	14 – 5 =
3.	15 – 10 =	23.	16 – 7 =
4.	18 – 10 =	24.	15 – 7 =
5.	12 – 2 =	25.	17 – 8 =
6.	11 – 10 =	26.	18 – 9 =
7.	17 – 13 =	27.	15 – 6 =
8.	20 – 10 =	28.	13 – 8 =
9.	14 – 11 =	29.	14 – 6 =
10.	16 – 12 =	30.	12 – 5 =
11.	11 – 3 =	31.	11 – 7 =
12.	13 – 2 =	32.	13 – 8 =
13.	14 – 2 =	33.	16 – 9 =
14.	13 – 4 =	34.	12 – 8 =
15.	12 – 3 =	35.	16 – 12 =
16.	11 – 4 =	36.	18 – 15 =
17.	12 – 5 =	37.	15 – 14 =
18.	14 – 5 =	38.	17 – 11 =
19	11 – 2 =	39.	19 – 13 =
20.	12 – 4 =	40.	20 – 12 =

Name _____   Date _____

1.	17 – 7 =		21.	16 – 7 =
2.	14 – 10 =		22.	17 – 8 =
3.	19 – 11 =		23.	18 – 7 =
4.	16 – 10 =		24.	14 – 6 =
5.	17 – 12 =		25.	17 – 8 =
6.	15 – 13 =		26.	12 – 8 =
7.	12 – 3 =		27.	14 – 7 =
8.	20 – 11 =		28.	15 – 8 =
9.	18 – 11 =		29.	13 – 5 =
10.	13 – 5 =		30.	16 – 8 =
11.	_____ = 11 – 2		31.	14 – 9 =
12.	_____ = 12 – 4		32.	15 – 6 =
13.	_____ = 13 – 5		33.	13 – 6 =
14.	_____ = 12 – 3		34.	_____ = 13 – 8
15.	_____ = 11 – 4		35.	_____ = 15 – 7
16.	_____ = 13 – 2		36.	_____ = 18 – 9
17.	_____ = 11 – 3		37.	_____ = 20 – 14
18.	17 – 8 =		38.	_____ = 20 – 7
19	14 – 6 =		39.	_____ = 20 – 11
20.	16 – 9 =		40.	_____ = 20 – 8

Lesson 1:   Sort and record data into a table using up to four categories; use
category counts to solve word problems.

© 2015 Great Minds®. eureka-math.org

71

Name _____   Date _____

1.	11 + 9 =	21.	13 − 7 =
2.	13 + 5 =	22.	11 − 8 =
3.	14 + 3 =	23.	15 − 6 =
4.	12 + 7 =	24.	12 + 7 =
5.	5 + 9 =	25.	14 + 3 =
6.	8 + 8 =	26.	8 + 12 =
7.	14 − 7 =	27.	5 + 7 =
8.	13 − 5 =	28.	8 + 9 =
9.	16 − 7 =	29.	7 + 5 =
10.	17 − 9 =	30.	13 − 6 =
11.	14 − 6 =	31.	14 − 8 =
12.	18 − 5 =	32.	12 − 9 =
13.	9 + 9 =	33.	11 − 3 =
14.	7 + 6 =	34.	14 − 5 =
15.	3 + 9 =	35.	13 − 8 =
16.	6 + 7 =	36.	8 + 5 =
17.	8 + 5 =	37.	4 + 7 =
18.	13 − 8 =	38.	7 + 8 =
19	16 − 9 =	39.	4 + 9 =
20.	14 − 8 =	40.	20 − 8 =

EUREKA MATH®

Lesson 1:   Sort and record data into a table using up to four categories; use category counts to solve word problems.

73

A

Number Correct: _____

Addition and Subtraction by 5

1.	0 + 5 =		23.	10 + 5 =	
2.	5 + 5 =		24.	15 + 5 =	
3.	10 + 5 =		25.	20 + 5 =	
4.	15 + 5 =		26.	25 + 5 =	
5.	20 + 5 =		27.	30 + 5 =	
6.	25 + 5 =		28.	35 + 5 =	
7.	30 + 5 =		29.	40 + 5 =	
8.	35 + 5 =		30.	45 + 5 =	
9.	40 + 5 =		31.	0 + 50 =	
10.	45 + 5 =		32.	50 + 50 =	
11.	50 – 5 =		33.	50 + 5 =	
12.	45 – 5 =		34.	55 + 5 =	
13.	40 – 5 =		35.	60 – 5 =	
14.	35 – 5 =		36.	55 – 5 =	
15.	30 – 5 =		37.	60 + 5 =	
16.	25 – 5 =		38.	65 + 5 =	
17.	20 – 5 =		39.	70 – 5 =	
18.	15 – 5 =		40.	65 – 5 =	
19.	10 – 5 =		41.	100 + 50 =	
20.	5 – 5 =		42.	150 + 50 =	
21.	5 + 0 =		43.	200 – 50 =	
22.	5 + 5 =		44.	150 – 50 =	

EUREKA MATH

Lesson 3: Draw and label a bar graph to represent data; relate the count scale to the number line.

© 2015 Great Minds®. eureka-math.org

75

B

Number Correct: _____

Improvement: _____

Addition and Subtraction by 5

1.	5 + 0 =	
2.	5 + 5 =	
3.	5 + 10 =	
4.	5 + 15 =	
5.	5 + 20 =	
6.	5 + 25 =	
7.	5 + 30 =	
8.	5 + 35 =	
9.	5 + 40 =	
10.	5 + 45 =	
11.	50 – 5 =	
12.	45 – 5 =	
13.	40 – 5 =	
14.	35 – 5 =	
15.	30 – 5 =	
16.	25 – 5 =	
17.	20 – 5 =	
18.	15 – 5 =	
19.	10 – 5 =	
20.	5 – 5 =	
21.	0 + 5 =	
22.	5 + 5 =	

23.	10 + 5 =	
24.	15 + 5 =	
25.	20 + 5 =	
26.	25 + 5 =	
27.	30 + 5 =	
28.	35 + 5 =	
29.	40 + 5 =	
30.	45 + 5 =	
31.	50 + 0 =	
32.	50 + 50 =	
33.	5 + 50 =	
34.	5 + 55 =	
35.	60 – 5 =	
36.	55 – 5 =	
37.	5 + 60 =	
38.	5 + 65 =	
39.	70 – 5 =	
40.	65 – 5 =	
41.	50 + 100 =	
42.	50 + 150 =	
43.	200 – 50 =	
44.	150 – 50 =	

EUREKA
MATH

Lesson 3: Draw and label a bar graph to represent data; relate the count scale to the number line.

© 2015 Great Minds®. eureka-math.org

77

A

Number Correct: _____

Skip-Counting by 5

1.	0, 5, ___	
2.	5, 10, ___	
3.	10, 15, ___	
4.	15, 20, ___	
5.	20, 25, ___	
6.	25, 30, ___	
7.	30, 35, ___	
8.	35, 40, ___	
9.	40, 45, ___	
10.	50, 45, ___	
11.	45, 40, ___	
12.	40, 35, ___	
13.	35, 30, ___	
14.	30, 25, ___	
15.	25, 20, ___	
16.	20, 15, ___	
17.	15, 10, ___	
18.	0, ___, 10	
19.	25, ___, 35	
20.	5, ___, 15	
21.	30, ___, 40	
22.	10, ___, 20	

23.	35, ___, 45	
24.	15, ___, 25	
25.	40, ___, 50	
26.	25, ___, 15	
27.	50, ___, 40	
28.	20, ___, 10	
29.	45, ___, 35	
30.	15, ___, 5	
31.	40, ___, 30	
32.	10, ___, 0	
33.	35, ___, 25	
34.	___, 10, 5	
35.	___, 35, 30	
36.	___, 15, 10	
37.	___, 40, 35	
38.	___, 20, 15	
39.	___, 45, 40	
40.	50, 55, ___	
41.	45, 50, ___	
42.	65, ___, 55	
43.	55, 60, ___	
44.	60, 65, ___	

B

Number Correct: _____

Skip-Counting by 5

Improvement: _____

1.	5, 10, ___	
2.	10, 15, ___	
3.	15, 20, ___	
4.	20, 25, ___	
5.	25, 30, ___	
6.	30, 35, ___	
7.	35, 40, ___	
8.	40, 45, ___	
9.	50, 45, ___	
10.	45, 40, ___	
11.	40, 35, ___	
12.	35, 30, ___	
13.	30, 25, ___	
14.	25, 20, ___	
15.	20, 15, ___	
16.	15, 10, ___	
17.	0, ___, 10	
18.	25, ___, 35	
19.	5, ___, 15	
20.	30, ___, 40	
21.	10, ___, 20	
22.	35, ___, 45	

23.	15, ___, 25	
24.	35, ___, 45	
25.	30, ___, 20	
26.	25, ___, 15	
27.	50, ___, 40	
28.	20, ___, 10	
29.	45, ___, 35	
30.	15, ___, 5	
31.	35, ___, 25	
32.	10, ___, 0	
33.	35, ___, 25	
34.	___, 15, 10	
35.	___, 40, 35	
36.	___, 20, 15	
37.	___, 45, 40	
38.	___, 10, 5	
39.	___, 35, 30	
40.	45, 50, ___	
41.	50, 55, ___	
42.	55, 60, ___	
43.	65, ___, 55	
44.	___, 60, 55	

A

Number Correct: _____

Subtraction Across a Ten

1.	10 – 3 =	
2.	11 – 3 =	
3.	12 – 3 =	
4.	10 – 2 =	
5.	11 – 2 =	
6.	10 – 5 =	
7.	11 – 5 =	
8.	12 – 5 =	
9.	14 – 5 =	
10.	10 – 4 =	
11.	11 – 4 =	
12.	12 – 4 =	
13.	13 – 4 =	
14.	10 – 7 =	
15.	11 – 7 =	
16.	12 – 7 =	
17.	15 – 7 =	
18.	10 – 6 =	
19.	11 – 6 =	
20.	12 – 6 =	
21.	14 – 6 =	
22.	10 – 9 =	

23.	11 – 9 =	
24.	12 – 9 =	
25.	17 – 9 =	
26.	10 – 8 =	
27.	11 – 8 =	
28.	12 – 8 =	
29.	16 – 8 =	
30.	10 – 6 =	
31.	13 – 6 =	
32.	15 – 6 =	
33.	10 – 7 =	
34.	13 – 7 =	
35.	14 – 7 =	
36.	16 – 7 =	
37.	10 – 8 =	
38.	13 – 8 =	
39.	14 – 8 =	
40.	17 – 8 =	
41.	10 – 9 =	
42.	13 – 9 =	
43.	14 – 9 =	
44.	18 – 9 =	

EUREKA MATH

Lesson 7: Solve word problems involving the total value of a group of coins.

83

B

Number Correct: _____

Subtraction Across a Ten

Improvement: _____

1.	10 – 2 =		23.	11 – 7 =	
2.	11 – 2 =		24.	12 – 7 =	
3.	10 – 4 =		25.	16 – 7 =	
4.	11 – 4 =		26.	10 – 9 =	
5.	12 – 4 =		27.	11 – 9 =	
6.	13 – 4 =		28.	12 – 9 =	
7.	10 – 3 =		29.	18 – 9 =	
8.	11 – 3 =		30.	10 – 5 =	
9.	12 – 3 =		31.	13 – 5 =	
10.	10 – 6 =		32.	10 – 6 =	
11.	11 – 6 =		33.	13 – 6 =	
12.	12 – 6 =		34.	14 – 6 =	
13.	15 – 6 =		35.	10 – 7 =	
14.	10 – 5 =		36.	13 – 7 =	
15.	11 – 5 =		37.	15 – 7 =	
16.	12 – 5 =		38.	10 – 8 =	
17.	14 – 5 =		39.	13 – 8 =	
18.	10 – 8 =		40.	14 – 8 =	
19.	11 – 8 =		41.	16 – 8 =	
20.	12 – 8 =		42.	10 – 9 =	
21.	17 – 8 =		43.	16 – 9 =	
22.	10 – 7 =		44.	17 – 9 =	

EUREKA MATH®

Lesson 7:   Solve word problems involving the total value of a group of coins.

85

A

Number Correct: _____

Adding Across a Ten

1.	9 + 2 =		23.	4 + 7 =	
2.	9 + 3 =		24.	4 + 8 =	
3.	9 + 4 =		25.	5 + 6 =	
4.	9 + 7 =		26.	5 + 7 =	
5.	7 + 9 =		27.	3 + 8 =	
6.	10 + 1 =		28.	3 + 9 =	
7.	10 + 2 =		29.	2 + 9 =	
8.	10 + 3 =		30.	5 + 10 =	
9.	10 + 8 =		31.	5 + 8 =	
10.	8 + 10 =		32.	9 + 6 =	
11.	8 + 3 =		33.	6 + 9 =	
12.	8 + 4 =		34.	7 + 6 =	
13.	8 + 5 =		35.	6 + 7 =	
14.	8 + 9 =		36.	8 + 6 =	
15.	9 + 8 =		37.	6 + 8 =	
16.	7 + 4 =		38.	8 + 7 =	
17.	10 + 5 =		39.	7 + 8 =	
18.	6 + 5 =		40.	6 + 6 =	
19.	7 + 5 =		41.	7 + 7 =	
20.	9 + 5 =		42.	8 + 8 =	
21.	5 + 9 =		43.	9 + 9 =	
22.	10 + 6 =		44.	4 + 9 =	

EUREKA MATH

Lesson 8: Solve word problems involving the total value of a group of bills.

87

© 2015 Great Minds®. eureka-math.org

B

Number Correct: _____

Improvement: _____

Adding Across a Ten

1.	10 + 1 =	
2.	10 + 2 =	
3.	10 + 3 =	
4.	10 + 9 =	
5.	9 + 10 =	
6.	9 + 2 =	
7.	9 + 3 =	
8.	9 + 4 =	
9.	9 + 8 =	
10.	8 + 9 =	
11.	8 + 3 =	
12.	8 + 4 =	
13.	8 + 5 =	
14.	8 + 7 =	
15.	7 + 8 =	
16.	7 + 4 =	
17.	10 + 4 =	
18.	6 + 5 =	
19.	7 + 5 =	
20.	9 + 5 =	
21.	5 + 9 =	
22.	10 + 8 =	

23.	5 + 6 =	
24.	5 + 7 =	
25.	4 + 7 =	
26.	4 + 8 =	
27.	4 + 10 =	
28.	3 + 8 =	
29.	3 + 9 =	
30.	2 + 9 =	
31.	5 + 8 =	
32.	7 + 6 =	
33.	6 + 7 =	
34.	8 + 6 =	
35.	6 + 8 =	
36.	9 + 6 =	
37.	6 + 9 =	
38.	9 + 7 =	
39.	7 + 9 =	
40.	6 + 6 =	
41.	7 + 7 =	
42.	8 + 8 =	
43.	9 + 9 =	
44.	4 + 9 =	

EUREKA MATH®

Lesson 8:   Solve word problems involving the total value of a group of bills.

© 2015 Great Minds®. eureka-math.org

89

A

Number Correct: _____

Subtraction from Teens

1.	11 – 10 =		23.	19 – 9 =	
2.	12 – 10 =		24.	15 – 6 =	
3.	13 – 10 =		25.	15 – 7 =	
4.	19 – 10 =		26.	15 – 9 =	
5.	11 – 1 =		27.	20 – 10 =	
6.	12 – 2 =		28.	14 – 5 =	
7.	13 – 3 =		29.	14 – 6 =	
8.	17 – 7 =		30.	14 – 7 =	
9.	11 – 2 =		31.	14 – 9 =	
10.	11 – 3 =		32.	15 – 5 =	
11.	11 – 4 =		33.	17 – 8 =	
12.	11 – 8 =		34.	17 – 9 =	
13.	18 – 8 =		35.	18 – 8 =	
14.	13 – 4 =		36.	16 – 7 =	
15.	13 – 5 =		37.	16 – 8 =	
16.	13 – 6 =		38.	16 – 9 =	
17.	13 – 8 =		39.	17 – 10 =	
18.	16 – 6 =		40.	12 – 8 =	
19.	12 – 3 =		41.	18 – 9 =	
20.	12 – 4 =		42.	11 – 9 =	
21.	12 – 5 =		43.	15 – 8 =	
22.	12 – 9 =		44.	13 – 7 =	

B

Number Correct: _____

Subtraction from Teens

Improvement: _____

1.	11 – 1 =		23.	16 – 6 =	
2.	12 – 2 =		24.	14 – 5 =	
3.	13 – 3 =		25.	14 – 6 =	
4.	18 – 8 =		26.	14 – 7 =	
5.	11 – 10 =		27.	14 – 9 =	
6.	12 – 10 =		28.	20 – 10 =	
7.	13 – 10 =		29.	15 – 6 =	
8.	18 – 10 =		30.	15 – 7 =	
9.	11 – 2 =		31.	15 – 9 =	
10.	11 – 3 =		32.	14 – 4 =	
11.	11 – 4 =		33.	16 – 7 =	
12.	11 – 7 =		34.	16 – 8 =	
13.	19 – 9 =		35.	16 – 9 =	
14.	12 – 3 =		36.	20 – 10 =	
15.	12 – 4 =		37.	17 – 8 =	
16.	12 – 5 =		38.	17 – 9 =	
17.	12 – 8 =		39.	16 – 10 =	
18.	17 – 7 =		40.	18 – 9 =	
19.	13 – 4 =		41.	12 – 9 =	
20.	13 – 5 =		42.	13 – 7 =	
21.	13 – 6 =		43.	11 – 8 =	
22.	13 – 9 =		44.	15 – 8 =	

A

Number Correct: _____

Adding Across a Ten

1.	9 + 2 =		23.	4 + 7 =	
2.	9 + 3 =		24.	4 + 8 =	
3.	9 + 4 =		25.	5 + 6 =	
4.	9 + 7 =		26.	5 + 7 =	
5.	7 + 9 =		27.	3 + 8 =	
6.	10 + 1 =		28.	3 + 9 =	
7.	10 + 2 =		29.	2 + 9 =	
8.	10 + 3 =		30.	5 + 10 =	
9.	10 + 8 =		31.	5 + 8 =	
10.	8 + 10 =		32.	9 + 6 =	
11.	8 + 3 =		33.	6 + 9 =	
12.	8 + 4 =		34.	7 + 6 =	
13.	8 + 5 =		35.	6 + 7 =	
14.	8 + 9 =		36.	8 + 6 =	
15.	9 + 8 =		37.	6 + 8 =	
16.	7 + 4 =		38.	8 + 7 =	
17.	10 + 5 =		39.	7 + 8 =	
18.	6 + 5 =		40.	6 + 6 =	
19.	7 + 5 =		41.	7 + 7 =	
20.	9 + 5 =		42.	8 + 8 =	
21.	5 + 9 =		43.	9 + 9 =	
22.	10 + 6 =		44.	4 + 9 =	

Lesson 12:   Solve word problems involving different ways to make change from $1.

95

EUREKA
MATH®

B

Adding Across a Ten

1.	10 + 1 =	
2.	10 + 2 =	
3.	10 + 3 =	
4.	10 + 9 =	
5.	9 + 10 =	
6.	9 + 2 =	
7.	9 + 3 =	
8.	9 + 4 =	
9.	9 + 8 =	
10.	8 + 9 =	
11.	8 + 3 =	
12.	8 + 4 =	
13.	8 + 5 =	
14.	8 + 7 =	
15.	7 + 8 =	
16.	7 + 4 =	
17.	10 + 4 =	
18.	6 + 5 =	
19.	7 + 5 =	
20.	9 + 5 =	
21.	5 + 9 =	
22.	10 + 8 =	

23.	5 + 6 =	
24.	5 + 7 =	
25.	4 + 7 =	
26.	4 + 8 =	
27.	4 + 10 =	
28.	3 + 8 =	
29.	3 + 9 =	
30.	2 + 9 =	
31.	5 + 8 =	
32.	7 + 6 =	
33.	6 + 7 =	
34.	8 + 6 =	
35.	6 + 8 =	
36.	9 + 6 =	
37.	6 + 9 =	
38.	9 + 7 =	
39.	7 + 9 =	
40.	6 + 6 =	
41.	7 + 7 =	
42.	8 + 8 =	
43.	9 + 9 =	
44.	4 + 9 =	

11 - 1	11 - 2
11 - 3	11 - 4
11 - 5	11 - 6
11 - 7	11 - 8
11 - 9	12 - 3

subtraction fact flash cards set 2

12 - 4	12 - 5
12 - 6	12 - 7
12 - 8	12 - 9
13 - 4	13 - 5
13 - 6	13 - 7

subtraction fact flash cards set 2

Lesson 14:   Connect measurement with physical units by using iteration with an inch tile to measure.

13 - 8	13 - 9
14 - 5	14 - 6
14 - 7	14 - 8
14 - 9	15 - 6
15 - 7	15 - 8

subtraction fact flash cards set 2

Lesson 14:   Connect measurement with physical units by using iteration with an inch tile to measure.

15 - 9	16 - 7
16 - 8	16 - 9
17 - 8	17 - 9
18 - 9	19 - 11
20 - 19	20 - 1

subtraction fact flash cards set 2

Lesson 14: Connect measurement with physical units by using iteration with an inch tile to measure.

105

20 - 18	20 - 2
20 - 17	20 - 3
20 - 16	20 - 4
20 - 15	20 - 5
20 - 14	20 - 6

subtraction fact flash cards set 2

Lesson 14:   Connect measurement with physical units by using iteration with an
inch tile to measure.

107

# 20 - 13	# 20 - 7
# 20 - 12	# 20 - 8
# 20 - 11	# 20 - 9
# 20 - 10	

subtraction fact flash cards set 2

Lesson 14:   Connect measurement with physical units by using iteration with an inch tile to measure.

© 2015 Great Minds®. eureka-math.org

109

A

Number Correct: _____

Adding and Subtracting by 2

1.	0 + 2 =	
2.	2 + 2 =	
3.	4 + 2 =	
4.	6 + 2 =	
5.	8 + 2 =	
6.	10 + 2 =	
7.	12 + 2 =	
8.	14 + 2 =	
9.	16 + 2 =	
10.	18 + 2 =	
11.	20 – 2 =	
12.	18 – 2 =	
13.	16 – 2 =	
14.	14 – 2 =	
15.	12 – 2 =	
16.	10 – 2 =	
17.	8 – 2 =	
18.	6 – 2 =	
19.	4 – 2 =	
20.	2 – 2 =	
21.	2 + 0 =	
22.	2 + 2 =	

23.	2 + 4 =	
24.	2 + 6 =	
25.	2 + 8 =	
26.	2 + 10 =	
27.	2 + 12 =	
28.	2 + 14 =	
29.	2 + 16 =	
30.	2 + 18 =	
31.	0 + 22 =	
32.	22 + 22 =	
33.	44 + 22 =	
34.	66 + 22 =	
35.	88 – 22 =	
36.	66 – 22 =	
37.	44 – 22 =	
38.	22 – 22 =	
39.	22 + 0 =	
40.	22 + 22 =	
41.	22 + 44 =	
42.	66 + 22 =	
43.	888 – 222 =	
44.	666 – 222 =	

Lesson 15:   Apply concepts to create inch rulers; measure lengths using inch rulers.

111

EUREKA MATH®

B

Number Correct: _____

Improvement: _____

Adding and Subtracting by 2

1.	2 + 0 =	
2.	2 + 2 =	
3.	2 + 4 =	
4.	2 + 6 =	
5.	2 + 8 =	
6.	2 + 10 =	
7.	2 + 12 =	
8.	2 + 14 =	
9.	2 + 16 =	
10.	2 + 18 =	
11.	20 – 2 =	
12.	18 – 2 =	
13.	16 – 2 =	
14.	14 – 2 =	
15.	12 – 2 =	
16.	10 – 2 =	
17.	8 – 2 =	
18.	6 – 2 =	
19.	4 – 2 =	
20.	2 – 2 =	
21.	0 + 2 =	
22.	2 + 2 =	

23.	4 + 2 =	
24.	6 + 2 =	
25.	8 + 2 =	
26.	10 + 2 =	
27.	12 + 2 =	
28.	14 + 2 =	
29.	16 + 2 =	
30.	18 + 2 =	
31.	0 + 22 =	
32.	22 + 22 =	
33.	22 + 44 =	
34.	66 + 22 =	
35.	88 – 22 =	
36.	66 – 22 =	
37.	44 – 22 =	
38.	22 – 22 =	
39.	22 + 0 =	
40.	22 + 22 =	
41.	22 + 44 =	
42.	66 + 22 =	
43.	666 – 222 =	
44.	888 – 222 =	

A

Number Correct: _____

Adding and Subtracting by 3

1.	$0 + 3 =$	
2.	$3 + 3 =$	
3.	$6 + 3 =$	
4.	$9 + 3 =$	
5.	$12 + 3 =$	
6.	$15 + 3 =$	
7.	$18 + 3 =$	
8.	$21 + 3 =$	
9.	$24 + 3 =$	
10.	$27 + 3 =$	
11.	$30 - 3 =$	
12.	$27 - 3 =$	
13.	$24 - 3 =$	
14.	$21 - 3 =$	
15.	$18 - 3 =$	
16.	$15 - 3 =$	
17.	$12 - 3 =$	
18.	$9 - 3 =$	
19.	$6 - 3 =$	
20.	$3 - 3 =$	
21.	$3 + 0 =$	
22.	$3 + 3 =$	

23.	$6 + 3 =$	
24.	$9 + 3 =$	
25.	$12 + 3 =$	
26.	$15 + 3 =$	
27.	$18 + 3 =$	
28.	$21 + 3 =$	
29.	$24 + 3 =$	
30.	$27 + 3 =$	
31.	$0 + 33 =$	
32.	$33 + 33 =$	
33.	$66 + 33 =$	
34.	$33 + 66 =$	
35.	$99 - 33 =$	
36.	$66 - 33 =$	
37.	$999 - 333 =$	
38.	$33 - 33 =$	
39.	$33 + 0 =$	
40.	$30 + 3 =$	
41.	$33 + 3 =$	
42.	$36 + 3 =$	
43.	$63 + 33 =$	
44.	$63 + 36 =$	

B

Number Correct: _____

Adding and Subtracting by 3

Improvement: _____

1.	3 + 0 =	
2.	3 + 3 =	
3.	3 + 6 =	
4.	3 + 9 =	
5.	3 + 12 =	
6.	3 + 15 =	
7.	3 + 18 =	
8.	3 + 21 =	
9.	3 + 24 =	
10.	3 + 27 =	
11.	30 – 3 =	
12.	27 – 3 =	
13.	24 – 3 =	
14.	21 – 3 =	
15.	18 – 3 =	
16.	15 – 3 =	
17.	12 – 3 =	
18.	9 – 3 =	
19.	6 – 3 =	
20.	3 – 3 =	
21.	0 + 3 =	
22.	3 + 3 =	

23.	6 + 3 =	
24.	9 + 3 =	
25.	12 + 3 =	
26.	15 + 3 =	
27.	18 + 3 =	
28.	21 + 3 =	
29.	24 + 3 =	
30.	27 + 3 =	
31.	0 + 33 =	
32.	33 + 33 =	
33.	33 + 66 =	
34.	66 + 33 =	
35.	99 – 33 =	
36.	66 – 33 =	
37.	999 – 333 =	
38.	33 – 33 =	
39.	33 + 0 =	
40.	30 + 3 =	
41.	33 + 3 =	
42.	36 + 3 =	
43.	36 + 33 =	
44.	36 + 63 =	

EUREKA MATH®

Lesson 16:   Measure various objects using inch rulers and yardsticks.

117

A

Number Correct: _____

Subtraction Patterns

1.	10 – 1 =	
2.	10 – 2 =	
3.	20 – 2 =	
4.	40 – 2 =	
5.	10 – 2 =	
6.	11 – 2 =	
7.	21 – 2 =	
8.	51 – 2 =	
9.	10 – 3 =	
10.	11 – 3 =	
11.	21 – 3 =	
12.	61 – 3 =	
13.	10 – 4 =	
14.	11 – 4 =	
15.	21 – 4 =	
16.	71 – 4 =	
17.	10 – 5 =	
18.	11 – 5 =	
19.	21 – 5 =	
20.	81 – 5 =	
21.	10 – 6 =	
22.	11 – 6 =	

23.	21 – 6 =	
24.	91 – 6 =	
25.	10 – 7 =	
26.	11 – 7 =	
27.	31 – 7 =	
28.	10 – 8 =	
29.	11 – 8 =	
30.	41 – 8 =	
31.	10 – 9 =	
32.	11 – 9 =	
33.	51 – 9 =	
34.	12 – 3 =	
35.	82 – 3 =	
36.	13 – 5 =	
37.	73 – 5 =	
38.	14 – 6 =	
39.	84 – 6 =	
40.	15 – 8 =	
41.	95 – 8 =	
42.	16 – 7 =	
43.	46 – 7 =	
44.	68 – 9 =	

EUREKA MATH®   Lesson 19:   Measure to compare the differences in lengths using inches, feet, and   119
                                   yards.

© 2015 Great Minds®. eureka-math.org

B

Number Correct: _____

Improvement: _____

Subtraction Patterns

1.	10 – 2 =	
2.	20 – 2 =	
3.	30 – 2 =	
4.	50 – 2 =	
5.	10 – 2 =	
6.	11 – 2 =	
7.	21 – 2 =	
8.	61 – 2 =	
9.	10 – 3 =	
10.	11 – 3 =	
11.	21 – 3 =	
12.	71 – 3 =	
13.	10 – 4 =	
14.	11 – 4 =	
15.	21 – 4 =	
16.	81 – 4 =	
17.	10 – 5 =	
18.	11 – 5 =	
19.	21 – 5 =	
20.	91 – 5 =	
21.	10 – 6 =	
22.	11 – 6 =	

23.	21 – 6 =	
24.	41 – 6 =	
25.	10 – 7 =	
26.	11 – 7 =	
27.	51 – 7 =	
28.	10 – 8 =	
29.	11 – 8 =	
30.	61 – 8 =	
31.	10 – 9 =	
32.	11 – 9 =	
33.	31 – 9 =	
34.	12 – 3 =	
35.	92 – 3 =	
36.	13 – 5 =	
37.	43 – 5 =	
38.	14 – 6 =	
39.	64 – 6 =	
40.	15 – 8 =	
41.	85 – 8 =	
42.	16 – 7 =	
43.	76 – 7 =	
44.	58 – 9 =	

EUREKA MATH

Lesson 19: Measure to compare the differences in lengths using inches, feet, and yards.

121

© 2015 Great Minds®. eureka-math.org

A

Number Correct: _____

Subtraction Patterns

1.	8 – 1 =	
2.	18 – 1 =	
3.	8 – 2 =	
4.	18 – 2 =	
5.	8 – 5 =	
6.	18 – 5 =	
7.	28 – 5 =	
8.	58 – 5 =	
9.	58 – 7 =	
10.	10 – 2 =	
11.	11 – 2 =	
12.	21 – 2 =	
13.	61 – 2 =	
14.	61 – 3 =	
15.	61 – 5 =	
16.	10 – 5 =	
17.	20 – 5 =	
18.	30 – 5 =	
19.	70 – 5 =	
20.	72 – 5 =	
21.	4 – 2 =	
22.	40 – 20 =	

23.	41 – 20 =	
24.	46 – 20 =	
25.	7 – 5 =	
26.	70 – 50 =	
27.	71 – 50 =	
28.	78 – 50 =	
29.	80 – 40 =	
30.	84 – 40 =	
31.	90 – 60 =	
32.	97 – 60 =	
33.	70 – 40 =	
34.	72 – 40 =	
35.	56 – 4 =	
36.	52 – 4 =	
37.	50 – 4 =	
38.	60 – 30 =	
39.	90 – 70 =	
40.	80 – 60 =	
41.	96 – 40 =	
42.	63 – 40 =	
43.	79 – 30 =	
44.	76 – 9 =	

B

Number Correct: _____

Improvement: _____

Subtraction Patterns

1.	7 – 1 =	
2.	17 – 1 =	
3.	7 – 2 =	
4.	17 – 2 =	
5.	7 – 5 =	
6.	17 – 5 =	
7.	27 – 5 =	
8.	57 – 5 =	
9.	57 – 6 =	
10.	10 – 5 =	
11.	11 – 5 =	
12.	21 – 5 =	
13.	61 – 5 =	
14.	61 – 4 =	
15.	61 – 2 =	
16.	10 – 2 =	
17.	20 – 2 =	
18.	30 – 2 =	
19.	70 – 2 =	
20.	71 – 2 =	
21.	5 – 2 =	
22.	50 – 20 =	

23.	51 – 20 =	
24.	56 – 20 =	
25.	8 – 5 =	
26.	80 – 50 =	
27.	81 – 50 =	
28.	87 – 50 =	
29.	60 – 30 =	
30.	64 – 30 =	
31.	80 – 60 =	
32.	85 – 60 =	
33.	70 – 30 =	
34.	72 – 30 =	
35.	76 – 4 =	
36.	72 – 4 =	
37.	70 – 4 =	
38.	80 – 40 =	
39.	90 – 60 =	
40.	60 – 40 =	
41.	93 – 40 =	
42.	67 – 40 =	
43.	78 – 30 =	
44.	56 – 9 =	

A

Number Correct: _____

Adding Across a Ten

1.	9 + 2 =		23.	4 + 7 =	
2.	9 + 3 =		24.	4 + 8 =	
3.	9 + 4 =		25.	5 + 6 =	
4.	9 + 7 =		26.	5 + 7 =	
5.	7 + 9 =		27.	3 + 8 =	
6.	10 + 1 =		28.	3 + 9 =	
7.	10 + 2 =		29.	2 + 9 =	
8.	10 + 3 =		30.	5 + 10 =	
9.	10 + 8 =		31.	5 + 8 =	
10.	8 + 10 =		32.	9 + 6 =	
11.	8 + 3 =		33.	6 + 9 =	
12.	8 + 4 =		34.	7 + 6 =	
13.	8 + 5 =		35.	6 + 7 =	
14.	8 + 9 =		36.	8 + 6 =	
15.	9 + 8 =		37.	6 + 8 =	
16.	7 + 4 =		38.	8 + 7 =	
17.	10 + 5 =		39.	7 + 8 =	
18.	6 + 5 =		40.	6 + 6 =	
19.	7 + 5 =		41.	7 + 7 =	
20.	9 + 5 =		42.	8 + 8 =	
21.	5 + 9 =		43.	9 + 9 =	
22.	10 + 6 =		44.	4 + 9 =	

EUREKA
MATH

Lesson 23:   Collect and record measurement data in a table; answer questions and
summarize the data set.

127

© 2015 Great Minds®. eureka-math.org

B

Number Correct: _____

Adding Across a Ten

Improvement: _____

1.	10 + 1 =		23.	5 + 6 =		
2.	10 + 2 =		24.	5 + 7 =		
3.	10 + 3 =		25.	4 + 7 =		
4.	10 + 9 =		26.	4 + 8 =		
5.	9 + 10 =		27.	4 + 10 =		
6.	9 + 2 =		28.	3 + 8 =		
7.	9 + 3 =		29.	3 + 9 =		
8.	9 + 4 =		30.	2 + 9 =		
9.	9 + 8 =		31.	5 + 8 =		
10.	8 + 9 =		32.	7 + 6 =		
11.	8 + 3 =		33.	6 + 7 =		
12.	8 + 4 =		34.	8 + 6 =		
13.	8 + 5 =		35.	6 + 8 =		
14.	8 + 7 =		36.	9 + 6 =		
15.	7 + 8 =		37.	6 + 9 =		
16.	7 + 4 =		38.	9 + 7 =		
17.	10 + 4 =		39.	7 + 9 =		
18.	6 + 5 =		40.	6 + 6 =		
19.	7 + 5 =		41.	7 + 7 =		
20.	9 + 5 =		42.	8 + 8 =		
21.	5 + 9 =		43.	9 + 9 =		
22.	10 + 8 =		44.	4 + 9 =		

EUREKA MATH®

Lesson 23: Collect and record measurement data in a table; answer questions and summarize the data set.

129

A

Number Correct: _____

Subtraction Patterns

1.	3 – 1 =		23.	8 – 7 =	
2.	13 – 1 =		24.	18 – 7 =	
3.	23 – 1 =		25.	58 – 7 =	
4.	53 – 1 =		26.	62 – 2 =	
5.	4 – 2 =		27.	9 – 8 =	
6.	14 – 2 =		28.	19 – 8 =	
7.	24 – 2 =		29.	29 – 8 =	
8.	64 – 2 =		30.	69 – 8 =	
9.	4 – 3 =		31.	7 – 3 =	
10.	14 – 3 =		32.	17 – 3 =	
11.	24 – 3 =		33.	77 – 3 =	
12.	74 – 3 =		34.	59 – 9 =	
13.	6 – 4 =		35.	9 – 7 =	
14.	16 – 4 =		36.	19 – 7 =	
15.	26 – 4 =		37.	89 – 7 =	
16.	96 – 4 =		38.	99 – 5 =	
17.	7 – 5 =		39.	78 – 6 =	
18.	17 – 5 =		40.	58 – 5 =	
19.	27 – 5 =		41.	39 – 7 =	
20.	47 – 5 =		42.	28 – 6 =	
21.	43 – 3 =		43.	49 – 4 =	
22.	87 – 7 =		44.	67 – 4 =	

B

Number Correct: _____

Subtraction Patterns

Improvement: _____

1.	2 – 1 =		23.	8 – 7 =	
2.	12 – 1 =		24.	18 – 7 =	
3.	22 – 1 =		25.	68 – 7 =	
4.	52 – 1 =		26.	32 – 2 =	
5.	5 – 2 =		27.	9 – 8 =	
6.	15 – 2 =		28.	19 – 8 =	
7.	25 – 2 =		29.	29 – 8 =	
8.	65 – 2 =		30.	79 – 8 =	
9.	4 – 3 =		31.	8 – 4 =	
10.	14 – 3 =		32.	18 – 4 =	
11.	24 – 3 =		33.	78 – 4 =	
12.	84 – 3 =		34.	89 – 9 =	
13.	7 – 4 =		35.	9 – 7 =	
14.	17 – 4 =		36.	19 – 7 =	
15.	27 – 4 =		37.	79 – 7 =	
16.	97 – 4 =		38.	89 – 5 =	
17.	6 – 5 =		39.	68 – 6 =	
18.	16 – 5 =		40.	48 – 5 =	
19.	26 – 5 =		41.	29 – 7 =	
20.	46 – 5 =		42.	38 – 6 =	
21.	23 – 3 =		43.	59 – 4 =	
22.	67 – 7 =		44.	77 – 4 =	

EUREKA MATH®

Lesson 24: Draw a line plot to represent the measurement data; relate the measurement scale to the number line.

133

Grade 2
Module 8

A

Number Correct: _____

Adding Across a Ten

1.	8 + 1 =	
2.	18 + 1 =	
3.	28 + 1 =	
4.	58 + 1 =	
5.	7 + 2 =	
6.	17 + 2 =	
7.	27 + 2 =	
8.	57 + 2 =	
9.	6 + 3 =	
10.	36 + 3 =	
11.	5 + 4 =	
12.	45 + 4 =	
13.	30 + 9 =	
14.	9 + 2 =	
15.	39 + 2 =	
16.	50 + 8 =	
17.	8 + 4 =	
18.	58 + 4 =	
19.	50 + 20 =	
20.	54 + 20 =	
21.	70 + 20 =	
22.	76 + 20 =	

23.	50 + 30 =	
24.	58 + 30 =	
25.	9 + 3 =	
26.	90 + 30 =	
27.	97 + 30 =	
28.	8 + 4 =	
29.	80 + 40 =	
30.	83 + 40 =	
31.	83 + 4 =	
32.	7 + 6 =	
33.	70 + 60 =	
34.	74 + 60 =	
35.	74 + 5 =	
36.	73 + 6 =	
37.	58 + 7 =	
38.	76 + 5 =	
39.	30 + 40 =	
40.	20 + 70 =	
41.	80 + 70 =	
42.	34 + 40 =	
43.	23 + 50 =	
44.	97 + 60 =	

B

Number Correct: _____

Adding Across a Ten

Improvement: _____

1.	7 + 1 =		23.	50 + 30 =	
2.	17 + 1 =		24.	57 + 30 =	
3.	27 + 1 =		25.	8 + 3 =	
4.	47 + 1 =		26.	80 + 30 =	
5.	6 + 2 =		27.	87 + 30 =	
6.	16 + 2 =		28.	9 + 4 =	
7.	26 + 2 =		29.	90 + 40 =	
8.	46 + 2 =		30.	93 + 40 =	
9.	5 + 3 =		31.	93 + 4 =	
10.	75 + 3 =		32.	8 + 6 =	
11.	5 + 4 =		33.	80 + 60 =	
12.	75 + 4 =		34.	84 + 60 =	
13.	40 + 9 =		35.	84 + 5 =	
14.	9 + 2 =		36.	83 + 6 =	
15.	49 + 2 =		37.	68 + 7 =	
16.	60 + 8 =		38.	86 + 5 =	
17.	8 + 4 =		39.	20 + 30 =	
18.	68 + 4 =		40.	30 + 60 =	
19.	50 + 20 =		41.	90 + 70 =	
20.	56 + 20 =		42.	36 + 40 =	
21.	70 + 20 =		43.	27 + 50 =	
22.	74 + 20 =		44.	94 + 70 =	

A

Make a Hundred to Add

1.	98 + 3 =		23.	99 + 12 =	
2.	98 + 4 =		24.	99 + 23 =	
3.	98 + 5 =		25.	99 + 34 =	
4.	98 + 8 =		26.	99 + 45 =	
5.	98 + 6 =		27.	99 + 56 =	
6.	98 + 9 =		28.	99 + 67 =	
7.	98 + 7 =		29.	99 + 78 =	
8.	99 + 2 =		30.	35 + 99 =	
9.	99 + 3 =		31.	45 + 98 =	
10.	99 + 4 =		32.	46 + 99 =	
11.	99 + 9 =		33.	56 + 98 =	
12.	99 + 6 =		34.	67 + 99 =	
13.	99 + 8 =		35.	77 + 98 =	
14.	99 + 5 =		36.	68 + 99 =	
15.	99 + 7 =		37.	78 + 98 =	
16.	98 + 13 =		38.	99 + 95 =	
17.	98 + 24 =		39.	93 + 99 =	
18.	98 + 35 =		40.	99 + 95 =	
19.	98 + 46 =		41.	94 + 99 =	
20.	98 + 57 =		42.	98 + 96 =	
21.	98 + 68 =		43.	94 + 98 =	
22.	98 + 79 =		44.	98 + 88 =	

EUREKA MATH®

Lesson 2: Build, identify, and analyze two-dimensional shapes with specified attributes.

© 2015 Great Minds®. eureka-math.org

141

B

Number Correct: _____

Make a Hundred to Add

Improvement: _____

1.	99 + 2 =		23.	98 + 13 =	
2.	99 + 3 =		24.	98 + 24 =	
3.	99 + 4 =		25.	98 + 35 =	
4.	99 + 8 =		26.	98 + 46 =	
5.	99 + 6 =		27.	98 + 57 =	
6.	99 + 9 =		28.	98 + 68 =	
7.	99 + 5 =		29.	98 + 79 =	
8.	99 + 7 =		30.	25 + 99 =	
9.	98 + 3 =		31.	35 + 98 =	
10.	98 + 4 =		32.	36 + 99 =	
11.	98 + 5 =		33.	46 + 98 =	
12.	98 + 9 =		34.	57 + 99 =	
13.	98 + 7 =		35.	67 + 98 =	
14.	98 + 8 =		36.	78 + 99 =	
15.	98 + 6 =		37.	88 + 98 =	
16.	99 + 12 =		38.	99 + 93 =	
17.	99 + 23 =		39.	95 + 99 =	
18.	99 + 34 =		40.	99 + 97 =	
19.	99 + 45 =		41.	92 + 99 =	
20.	99 + 56 =		42.	98 + 94 =	
21.	99 + 67 =		43.	96 + 98 =	
22.	99 + 78 =		44.	98 + 86 =	

Lesson 2:   Build, identify, and analyze two-dimensional shapes with specified attributes.

143

© 2015 Great Minds®. eureka-math.org

Name _____   Date _____

1.	10 + 9 =	21.	3 + 9 =
2.	10 + 1 =	22.	4 + 8 =
3.	11 + 2 =	23.	5 + 9 =
4.	13 + 6 =	24.	8 + 8 =
5.	15 + 5 =	25.	7 + 5 =
6.	14 + 3 =	26.	5 + 8 =
7.	13 + 5 =	27.	8 + 3 =
8.	12 + 4 =	28.	6 + 8 =
9.	16 + 2 =	29.	4 + 6 =
10.	18 + 1 =	30.	7 + 6 =
11.	11 + 7 =	31.	7 + 4 =
12.	13 + 4 =	32.	7 + 9 =
13.	14 + 5 =	33.	7 + 7 =
14.	9 + 4 =	34.	8 + 6 =
15.	9 + 2 =	35.	6 + 9 =
16.	9 + 9 =	36.	8 + 5 =
17.	6 + 9 =	37.	4 + 7 =
18.	8 + 9 =	38.	3 + 9 =
19.	7 + 8 =	39.	8 + 6 =
20.	8 + 8 =	40.	9 + 4 =

Lesson 3: Use attributes to draw different polygons including triangles, quadrilaterals, pentagons, and hexagons.

Name _____   Date _____

1.	10 + 8 =	21.	5 + 8 =
2.	4 + 10 =	22.	6 + 7 =
3.	9 + 10 =	23.	_____ + 4 = 12
4.	11 + 5 =	24.	_____ + 7 = 13
5.	13 + 3 =	25.	6 + _____ = 14
6.	12 + 4 =	26.	7 + _____ = 15
7.	16 + 3 =	27.	_____ = 9 + 8
8.	15 + _____ = 19	28.	_____ = 7 + 5
9.	18 + _____ = 20	29.	_____ = 4 + 8
10.	13 + 5 =	30.	3 + 9 =
11.	_____ = 4 + 16	31.	6 + 7 =
12.	_____ = 6 + 12	32.	8 + _____ = 13
13.	_____ = 14 + 6	33.	_____ = 7 + 9
14.	9 + 3 =	34.	6 + 6 =
15.	7 + 9 =	35.	_____ = 7 + 5
16.	_____ + 4 = 11	36.	_____ = 4 + 8
17.	_____ + 6 = 13	37.	20 = 13 + _____
18.	_____ + 5 = 12	38.	18 = _____ + 9
19.	_____ + 8 = 14	39.	16 = _____ + 7
20.	_____ + 9 = 15	40.	20 = 9 + _____

Lesson 3:   Use attributes to draw different polygons including triangles, quadrilaterals, pentagons, and hexagons.

Name _____   Date _____

1.	19 – 9 =	21.	15 – 7 =
2.	19 – 11 =	22.	18 – 9 =
3.	17 – 10 =	23.	16 – 8 =
4.	12 – 2 =	24.	15 – 6 =
5.	15 – 12 =	25.	17 – 8 =
6.	18 – 10 =	26.	14 – 6 =
7.	17 – 5 =	27.	16 – 9 =
8.	20 – 9 =	28.	13 – 8 =
9.	14 – 4 =	29.	12 – 5 =
10.	16 – 13 =	30.	19 – 8 =
11.	11 – 2 =	31.	17 – 9 =
12.	12 – 3 =	32.	16 – 7 =
13.	14 – 2 =	33.	14 – 8 =
14.	13 – 4 =	34.	15 – 9 =
15.	11 – 3 =	35.	13 – 7 =
16.	12 – 4 =	36.	12 – 8 =
17.	13 – 2 =	37.	15 – 8 =
18.	14 – 5 =	38.	14 – 9 =
19.	11 – 4 =	39.	12 – 7 =
20.	12 – 5 =	40.	11 – 9 =

Name _____   Date _____

1.	12 – 3 =	21.	13 – 7 =
2.	13 – 5 =	22.	15 – 9 =
3.	11 – 2 =	23.	18 – 7 =
4.	12 – 5 =	24.	14 – 7 =
5.	13 – 4 =	25.	17 – 9 =
6.	13 – 2 =	26.	12 – 9 =
7.	11 – 4 =	27.	13 – 6 =
8.	12 – 6 =	28.	15 – 7 =
9.	11 – 3 =	29.	16 – 8 =
10.	13 – 6 =	30.	12 – 6 =
11.	_____ = 11 – 9	31.	_____ = 13 – 9
12.	_____ = 13 – 8	32.	_____ = 17 – 8
13.	_____ = 12 – 7	33.	_____ = 14 – 9
14.	_____ = 11 – 6	34.	_____ = 13 – 5
15.	_____ = 13 – 9	35.	_____ = 15 – 8
16.	_____ = 14 – 8	36.	_____ = 18 – 9
17.	_____ = 11 – 7	37.	_____ = 16 – 7
18.	_____ = 15 – 6	38.	_____ = 20 – 12
19.	_____ = 16 – 9	39.	_____ = 20 – 6
20.	_____ = 12 – 8	40.	_____ = 20 – 17

Lesson 3:   Use attributes to draw different polygons including triangles, quadrilaterals, pentagons, and hexagons.

© 2015 Great Minds®. eureka-math.org

151

Name _____   Date _____

1.	13 – 4 =	21.	8 + 4 =
2.	15 – 8 =	22.	6 + 7 =
3.	19 – 5 =	23.	9 + 9 =
4.	11 – 7 =	24.	12 – 6 =
5.	9 + 6 =	25.	16 – 7 =
6.	7 + 8 =	26.	13 – 5 =
7.	4 + 7 =	27.	11 – 8 =
8.	13 + 6 =	28.	7 + 9 =
9.	12 – 8 =	29.	5 + 7 =
10.	17 – 9 =	30.	8 + 7 =
11.	14 – 6 =	31.	9 + 8 =
12.	16 – 7 =	32.	11 + 9 =
13.	6 + 8 =	33.	12 – 3 =
14.	7 + 6 =	34.	14 – 5 =
15.	4 + 9 =	35.	20 – 13 =
16.	5 + 7 =	36.	8 – 5 =
17.	9 – 5 =	37.	7 + 4 =
18.	13 – 7 =	38.	13 + 5 =
19.	16 – 9 =	39.	7 + 9 =
20.	14 – 8 =	40.	8 + 11 =

Hundreds	Tens	Ones

Workspace:

hundreds place value chart

Lesson 3:   Use attributes to draw different polygons including triangles,
quadrilaterals, pentagons, and hexagons.

155

© 2015 Great Minds®. eureka-math.org

A

Number Correct: _____

Subtraction Patterns

1.	8 – 1 =		23.	41 – 20 =	
2.	18 – 1 =		24.	46 – 20 =	
3.	8 – 2 =		25.	7 – 5 =	
4.	18 – 2 =		26.	70 – 50 =	
5.	8 – 5 =		27.	71 – 50 =	
6.	18 – 5 =		28.	78 – 50 =	
7.	28 – 5 =		29.	80 – 40 =	
8.	58 – 5 =		30.	84 – 40 =	
9.	58 – 7 =		31.	90 – 60 =	
10.	10 – 2 =		32.	97 – 60 =	
11.	11 – 2 =		33.	70 – 40 =	
12.	21 – 2 =		34.	72 – 40 =	
13.	61 – 2 =		35.	56 – 4 =	
14.	61 – 3 =		36.	52 – 4 =	
15.	61 – 5 =		37.	50 – 4 =	
16.	10 – 5 =		38.	60 – 30 =	
17.	20 – 5 =		39.	90 – 70 =	
18.	30 – 5 =		40.	80 – 60 =	
19.	70 – 5 =		41.	96 – 40 =	
20.	72 – 5 =		42.	63 – 40 =	
21.	4 – 2 =		43.	79 – 30 =	
22.	40 – 20 =		44.	76 – 9 =	

EUREKA MATH®

Lesson 5:   Relate the square to the cube, and describe the cube based on attributes.

157

B

Number Correct: _____

Subtraction Patterns

Improvement: _____

1.	7 – 1 =		23.	51 – 20 =	
2.	17 – 1 =		24.	56 – 20 =	
3.	7 – 2 =		25.	8 – 5 =	
4.	17 – 2 =		26.	80 – 50 =	
5.	7 – 5 =		27.	81 – 50 =	
6.	17 – 5 =		28.	87 – 50 =	
7.	27 – 5 =		29.	60 – 30 =	
8.	57 – 5 =		30.	64 – 30 =	
9.	57 – 6 =		31.	80 – 60 =	
10.	10 – 5 =		32.	85 – 60 =	
11.	11 – 5 =		33.	70 – 30 =	
12.	21 – 5 =		34.	72 – 30 =	
13.	61 – 5 =		35.	76 – 4 =	
14.	61 – 4 =		36.	72 – 4 =	
15.	61 – 2 =		37.	70 – 4 =	
16.	10 – 2 =		38.	80 – 40 =	
17.	20 – 2 =		39.	90 – 60 =	
18.	30 – 2 =		40.	60 – 40 =	
19.	70 – 2 =		41.	93 – 40 =	
20.	71 – 2 =		42.	67 – 40 =	
21.	5 – 2 =		43.	78 – 30 =	
22.	50 – 20 =		44.	56 – 9 =	

Lesson 5:   Relate the square to the cube, and describe the cube based on attributes.

© 2015 Great Minds®. eureka-math.org

A

Number Correct: _____

Addition and Subtraction Patterns

1.	8 + 3 =	
2.	11 – 3 =	
3.	9 + 2 =	
4.	11 – 2 =	
5.	6 + 5 =	
6.	11 – 6 =	
7.	7 + 4 =	
8.	11 – 7 =	
9.	8 + 4 =	
10.	12 – 4 =	
11.	9 + 3 =	
12.	12 – 3 =	
13.	7 + 5 =	
14.	12 – 7 =	
15.	6 + 6 =	
16.	12 – 6 =	
17.	8 + 6 =	
18.	14 – 8 =	
19.	9 + 4 =	
20.	13 – 9 =	
21.	8 + 7 =	
22.	15 – 8 =	

23.	8 + 8 =	
24.	16 – 8 =	
25.	9 + 6 =	
26.	15 – 9 =	
27.	9 + 9 =	
28.	18 – 9 =	
29.	7 + 7 =	
30.	14 – 7 =	
31.	8 + 9 =	
32.	17 – 8 =	
33.	7 + 9 =	
34.	16 – 7 =	
35.	19 – 6 =	
36.	6 + 7 =	
37.	17 – 6 =	
38.	11 – 7 =	
39.	7 + 6 =	
40.	13 – 7 =	
41.	19 – 7 =	
42.	3 + 8 =	
43.	5 + 8 =	
44.	18 – 5 =	

EUREKA MATH

Lesson 6: Combine shapes to create a composite shape; create a new shape from composite shapes.

161

© 2015 Great Minds®. eureka-math.org

B

Number Correct: _____

Addition and Subtraction Patterns

Improvement: _____

1.	9 + 2 =	
2.	11 – 2 =	
3.	8 + 3 =	
4.	11 – 3 =	
5.	7 + 4 =	
6.	11 – 7 =	
7.	6 + 5 =	
8.	11 – 6 =	
9.	9 + 3 =	
10.	12 – 3 =	
11.	8 + 4 =	
12.	12 – 4 =	
13.	7 + 5 =	
14.	12 – 5 =	
15.	6 + 6 =	
16.	12 – 6 =	
17.	9 + 4 =	
18.	13 – 4 =	
19.	8 + 6 =	
20.	14 – 8 =	
21.	7 + 8 =	
22.	15 – 7 =	

23.	9 + 6 =	
24.	15 – 9 =	
25.	8 + 8 =	
26.	16 – 8 =	
27.	7 + 7 =	
28.	14 – 7 =	
29.	9 + 9 =	
30.	18 – 9 =	
31.	7 + 9 =	
32.	16 – 9 =	
33.	8 + 9 =	
34.	17 – 9 =	
35.	19 – 7 =	
36.	5 + 8 =	
37.	18 – 5 =	
38.	13 – 8 =	
39.	6 + 7 =	
40.	13 – 6 =	
41.	19 – 6 =	
42.	3 + 9 =	
43.	6 + 9 =	
44.	18 – 6 =	

EUREKA MATH®

Lesson 6: Combine shapes to create a composite shape; create a new shape from composite shapes.

163

A

Number Correct: _____

Subtraction Patterns

1.	5 – 1 =	
2.	15 – 1 =	
3.	25 – 1 =	
4.	75 – 1 =	
5.	5 – 2 =	
6.	15 – 2 =	
7.	25 – 2 =	
8.	75 – 2 =	
9.	4 – 1 =	
10.	40 – 10 =	
11.	43 – 10 =	
12.	43 – 20 =	
13.	43 – 21 =	
14.	43 – 23 =	
15.	12 – 2 =	
16.	62 – 2 =	
17.	62 – 12 =	
18.	18 – 8 =	
19.	78 – 8 =	
20.	78 – 18 =	
21.	41 – 11 =	
22.	92 – 12 =	

23.	10 – 2 =	
24.	11 – 2 =	
25.	21 – 2 =	
26.	31 – 2 =	
27.	51 – 2 =	
28.	51 – 12 =	
29.	10 – 5 =	
30.	11 – 5 =	
31.	12 – 5 =	
32.	22 – 5 =	
33.	32 – 5 =	
34.	62 – 5 =	
35.	62 – 15 =	
36.	72 – 15 =	
37.	82 – 15 =	
38.	32 – 15 =	
39.	10 – 9 =	
40.	11 – 9 =	
41.	51 – 9 =	
42.	51 – 10 =	
43.	51 – 19 =	
44.	65 – 46 =	

EUREKA MATH

Lesson 9: Partition circles and rectangles into equal parts, and describe those parts as halves, thirds, or fourths.

165

© 2015 Great Minds®. eureka-math.org

B

Number Correct: _____

Improvement: _____

Subtraction Patterns

1.	4 – 1 =		23.	10 – 5 =	
2.	14 – 1 =		24.	11 – 5 =	
3.	24 – 1 =		25.	21 – 5 =	
4.	74 – 1 =		26.	31 – 5 =	
5.	5 – 3 =		27.	51 – 5 =	
6.	15 – 3 =		28.	51 – 15 =	
7.	25 – 3 =		29.	10 – 9 =	
8.	75 – 3 =		30.	11 – 9 =	
9.	3 – 1 =		31.	12 – 9 =	
10.	30 – 10 =		32.	22 – 9 =	
11.	32 – 10 =		33.	32 – 9 =	
12.	32 – 20 =		34.	62 – 9 =	
13.	32 – 21 =		35.	62 – 19 =	
14.	32 – 22 =		36.	72 – 19 =	
15.	15 – 5 =		37.	82 – 19 =	
16.	65 – 5 =		38.	32 – 19 =	
17.	65 – 15 =		39.	10 – 2 =	
18.	16 – 6 =		40.	11 – 2 =	
19.	76 – 6 =		41.	51 – 2 =	
20.	76 – 16 =		42.	51 – 10 =	
21.	51 – 11 =		43.	51 – 12 =	
22.	82 – 12 =		44.	95 – 76 =	

EUREKA MATH®

Lesson 9: Partition circles and rectangles into equal parts, and describe those parts as halves, thirds, or fourths.

167

A

Number Correct: _____

Addition Patterns

1.	8 + 2 =		23.	18 + 6 =	
2.	18 + 2 =		24.	28 + 6 =	
3.	38 + 2 =		25.	16 + 8 =	
4.	7 + 3 =		26.	26 + 8 =	
5.	17 + 3 =		27.	18 + 7 =	
6.	37 + 3 =		28.	18 + 8 =	
7.	8 + 3 =		29.	28 + 7 =	
8.	18 + 3 =		30.	28 + 8 =	
9.	28 + 3 =		31.	15 + 9 =	
10.	6 + 5 =		32.	16 + 9 =	
11.	16 + 5 =		33.	25 + 9 =	
12.	26 + 5 =		34.	26 + 9 =	
13.	18 + 4 =		35.	14 + 7 =	
14.	28 + 4 =		36.	16 + 6 =	
15.	16 + 6 =		37.	15 + 8 =	
16.	26 + 6 =		38.	23 + 8 =	
17.	18 + 5 =		39.	25 + 7 =	
18.	28 + 5 =		40.	15 + 7 =	
19.	16 + 7 =		41.	24 + 7 =	
20.	26 + 7 =		42.	14 + 9 =	
21.	19 + 2 =		43.	19 + 8 =	
22.	17 + 4 =		44.	28 + 9 =	

EUREKA
MATH

Lesson 10: Partition circles and rectangles into equal parts, and describe those parts as halves, thirds, or fourths.

169

© 2015 Great Minds®. eureka-math.org

B

Number Correct: _____

Addition Patterns

Improvement: _____

1.	9 + 1 =	
2.	19 + 1 =	
3.	39 + 1 =	
4.	6 + 4 =	
5.	16 + 4 =	
6.	36 + 4 =	
7.	9 + 2 =	
8.	19 + 2 =	
9.	29 + 2 =	
10.	7 + 4 =	
11.	17 + 4 =	
12.	27 + 4 =	
13.	19 + 3 =	
14.	29 + 3 =	
15.	17 + 5 =	
16.	27 + 5 =	
17.	19 + 4 =	
18.	29 + 4 =	
19.	17 + 6 =	
20.	27 + 6 =	
21.	18 + 3 =	
22.	26 + 5 =	

23.	19 + 5 =	
24.	29 + 5 =	
25.	17 + 7 =	
26.	27 + 7 =	
27.	19 + 6 =	
28.	19 + 7 =	
29.	29 + 6 =	
30.	29 + 7 =	
31.	17 + 8 =	
32.	17 + 9 =	
33.	27 + 8 =	
34.	27 + 9 =	
35.	12 + 9 =	
36.	14 + 8 =	
37.	16 + 7 =	
38.	28 + 6 =	
39.	26 + 8 =	
40.	24 + 8 =	
41.	13 + 8 =	
42.	24 + 9 =	
43.	29 + 8 =	
44.	18 + 9 =	

EUREKA MATH®

Lesson 10: Partition circles and rectangles into equal parts, and describe those parts as halves, thirds, or fourths.

A

Number Correct: _____

Adding and Subtracting by 5

1.	0 + 5 =	
2.	5 + 5 =	
3.	10 + 5 =	
4.	15 + 5 =	
5.	20 + 5 =	
6.	25 + 5 =	
7.	30 + 5 =	
8.	35 + 5 =	
9.	40 + 5 =	
10.	45 + 5 =	
11.	50 – 5 =	
12.	45 – 5 =	
13.	40 – 5 =	
14.	35 – 5 =	
15.	30 – 5 =	
16.	25 – 5 =	
17.	20 – 5 =	
18.	15 – 5 =	
19.	10 – 5 =	
20.	5 – 5 =	
21.	5 + 0 =	
22.	5 + 5 =	

23.	10 + 5 =	
24.	15 + 5 =	
25.	20 + 5 =	
26.	25 + 5 =	
27.	30 + 5 =	
28.	35 + 5 =	
29.	40 + 5 =	
30.	45 + 5 =	
31.	0 + 50 =	
32.	50 + 50 =	
33.	50 + 5 =	
34.	55 + 5 =	
35.	60 – 5 =	
36.	55 – 5 =	
37.	60 + 5 =	
38.	65 + 5 =	
39.	70 – 5 =	
40.	65 – 5 =	
41.	100 + 50 =	
42.	150 + 50 =	
43.	200 – 50 =	
44.	150 – 50 =	

B

Number Correct: _____

Adding and Subtracting by 5

Improvement: _____

1.	5 + 0 =		23.	10 + 5 =	
2.	5 + 5 =		24.	15 + 5 =	
3.	5 + 10 =		25.	20 + 5 =	
4.	5 + 15 =		26.	25 + 5 =	
5.	5 + 20 =		27.	30 + 5 =	
6.	5 + 25 =		28.	35 + 5 =	
7.	5 + 30 =		29.	40 + 5 =	
8.	5 + 35 =		30.	45 + 5 =	
9.	5 + 40 =		31.	50 + 0 =	
10.	5 + 45 =		32.	50 + 50 =	
11.	50 − 5 =		33.	5 + 50 =	
12.	45 − 5 =		34.	5 + 55 =	
13.	40 − 5 =		35.	60 − 5 =	
14.	35 − 5 =		36.	55 − 5 =	
15.	30 − 5 =		37.	5 + 60 =	
16.	25 − 5 =		38.	5 + 65 =	
17.	20 − 5 =		39.	70 − 5 =	
18.	15 − 5 =		40.	65 − 5 =	
19.	10 − 5 =		41.	50 + 100 =	
20.	5 − 5 =		42.	50 + 150 =	
21.	0 + 5 =		43.	200 − 50 =	
22.	5 + 5 =		44.	150 − 50 =	

Credits

Great Minds® has made every effort to obtain permission for the reprinting of all copyrighted material. If any owner of copyrighted material is not acknowledged herein, please contact Great Minds for proper acknowledgment in all future editions and reprints of this module.